Richard Anthony Proctor

Flowers of the Sky

Richard Anthony Proctor

Flowers of the Sky

ISBN/EAN: 9783337038731

Printed in Europe, USA, Canada, Australia, Japan

Cover: Foto ©berggeist007 / pixelio.de

More available books at **www.hansebooks.com**

SKY

By RICHARD A. PROCTOR

AUTHOR OF "THE EXPANSE OF HEAVEN," "THE INFINITIES AROUND US,"
"THE UNIVERSE OF STARS," "THE SUN," "THE MOON,"
ETC., ETC.

WITH FIFTY-FOUR ILLUSTRATIONS

New York
A. C. ARMSTRONG AND SON
714, BROADWAY

"Spirit of nature! here,
In this interminable wilderness
Of worlds, at whose immensity
　Even soaring fancy staggers,
　Here is thy fitting temple.
　　Yet not the lightest leaf
That quivers to the passing breeze
　Is less instinct with thee."—*Shelley*.

CONTENTS.

		PAGE
I.	LIGHT	1
II.	SPACE	17
III.	THE INFINITELY MINUTE	31
IV.	THE MYSTERY OF GRAVITY	43
V.	THE END OF MANY WORLDS	56
VI.	THE AURORA BOREALIS	91
VII.	THE LUNAR HALO	114
VIII.	MOONLIGHT	125
IX.	THE PLANET MARS	149
X.	THE PLANET JUPITER	191
XI.	THE RINGED PLANET SATURN	215
XII.	FANCIED FIGURES AMONG THE STARS	236
XIII.	TRANSITS OF VENUS	273

I.

LIGHT.

" What soul was his, when, from the naked top
 Of some bold headland, he beheld the sun
 Rise up and bathe the world in light ! He looked—
 Ocean and earth, the solid frame of earth
 And ocean's liquid mass, beneath him lay
 In gladness and deep joy."

E live in a mighty ocean whose waves are ever rushing hither and thither, always according to law, with velocity inconceivable, almost immeasurable. These waves lave the shore of that island of space which is our home, travelling to it from remotest regions, and making known to us all that we know of what lies outside our small abode. We call these waves, or rather their effects, by the name of Light. We recognise in light—

"offspring of Heav'n's first-born
And of th' Eternal co-eternal beam"—

the antecedent of all else that exists in the universe;

or, as Sir John Herschel said, "the superior in point of rank and conception to all other products or results of creative power in the physical world. It is light which alone can give, and does give us, the assurance of a uniform and all-pervading energy—a mechanism almost beyond conception, complex, minute, and powerful, by which that influence, or rather that movement, is propagated. Our evidence of the existence of gravitation fails us beyond the region of the double stars, or leaves at best only a presumption amounting to moral conviction in its favour. But the argument for a unity of design and action afforded by light stands unweakened by distance, and is co-extensive with the universe itself."

What, then, is light? What is that mysterious movement of some essence pervading all space, whereby, from remotest depths, news is brought to us, after journeys lasting many years, though space is traversed at a rate exceeding more than ten million times the velocity of the swiftest express train?

Light is in reality the result of undulations in what is called the ether of space, a perfectly transparent, almost perfectly elastic medium, occupying not only void space, but flowing as freely through the densest solids as the summer breeze flows through the forest trees. The waves of light cannot in this way pass through solid or liquid, or even aerial bodies, but either they are sooner or later brought to rest, or else they are more or less gradually deflected; just as the waves which traverse the ocean come to their end, or are deflected, when they meet the shore or shallows near the shore.

All light, however, has its real origin, not in the ethereal ocean itself, but in the movements of the minute particles of which all forms of matter known to us are composed. A tiny atom, far too small to be perceived with a microscope, even though one should be made ten thousand times more powerful than any yet constructed, when set in rapid vibration, raises minute waves in the ethereal ocean, just as a small body, vibrating on the surface of a sheet of water, would generate waves there. And as the water-waves would travel radially away from the place of their birth, so do the light-waves generated by the vibrations of one of the atoms composing a luminous body radiate forth in all directions through the ethereal ocean until, encountering some obstacle, they are sent (reduced in size) in a new direction.

In some luminous bodies there are atoms vibrating in many different periods (all very small) so as to cause light-waves of many different kinds to proceed from the body. In other cases the atoms all vibrate at one rate, or at two or three or some definite number of rates, so that only light-waves of certain kinds proceed from the body. But in all cases these light-waves only cause us to see the body when they flow in through the pupil of the eye, and falling upon the retina (or the choroid membrane, or whatever part of the eye it may be which finally receives the waves), convey to the optic nerve, and thence to the brain, the information that such and such a body, so coloured, so shaped, so moving, exists towards that direction from which the light-waves seem to come. The body so *seen*, as we call it, may be the original source of light, or may be a body on which light has been reflected to us.

It is in this way that we receive information from light-waves. It will be conceived how minute they must be, how perfectly they must retain their separate character, multitudinous though they are, in traversing the ether (even when that ether is clogged by the gross matter of our ordinary air), if we remember how through the tiny eye-pupil we often receive light-waves telling us of all the details, all the varieties of colour and brightness, all the movements in a rich landscape.

Even more startling are the thoughts suggested by a view of the starlit heavens. From hundreds of suns at once the light-waves which have traversed varying but all enormous distances pour in upon the small circle of the eye-pupil, waves of many kinds coming in together from each sun. The waves which thus reach the eye from one bright star have been but a few years upon their journey; all that time they have been traversing an ocean swept in every part by untold millions of other waves, and yet they arrive as perfect in order and regularity as rollers which have traversed a wide sea pour in upon a level shore. From another star, as bright as the first, they have been years in travelling; from some among the fainter stars, hundreds, perhaps thousands of years. Yet still they flow on, each order of waves in perfect uniformity as when they first left their parent sun.

But even this is not all. Among the waves which reach the eye many, nay, most, are so small that ordinary vision cannot perceive their action. Take, however, a telescope, and so gather them together as to intensify this action, and they are rendered perceptible, just as the unnoticed heaving of ocean becomes a manifest wave-motion when it reaches a regularly narrowing inlet. Thus, from stars so remote that their light has required thousands, or, even in some cases, perhaps, hundreds of thousands of years in reaching us, the light-waves flow steadily in upon us. So small are these waves, that the breadth of from forty to sixty thousand of them would occupy but a single inch. Through every point in space waves from all the hundred millions of stars are at all times simultaneously rushing at the rate of one hundred and eighty-five thousand miles in every second of time: yet they travel on altogether undisturbed, and each tells its story as distinctly as though the ether had conveyed no other message, and that message but for a short distance.

It would be difficult to say which thought, considered in its real significance, is more striking,—the thought of what is done for us by light regarded as a terrestrial phenomenon, or the thought of what light is doing, and has done, in presenting to us a view of the starlit heavens.

When the sun rises in splendour above the eastern horizon, tinting the sky with varied colours, lighting up the clouds which till then have been but dark patches on the heavens, bringing out the colours of hill and dale, rock and river, fields and woods, the heart gladdens at the spectacle. A pleasing melancholy falls on us as the light fades away at eventide, tint after tint vanishing, until at length the gloom of night enshrouds all. The full splendour of mid-day, the chastened splendour of a moonlit night, and the glory of the heavens when "all the stars shine, and the shepherd gladdens in his heart," stir the soul in like manner; and it might seem to many that to analyse these glories, to explain their scientific meaning, would be to rob the mind of the pleasure it had before found in such scenes. Many would be disposed to think that a purer enjoyment is expressed by Augustine than any student of science could find in the wonders of light, in those words in which he expresses

Fig. 1.—Sunrise on the Righi.

his sense of the loveliness of fair forms and brilliant colours. "For light, queen of the colours," he says, "bathing all I can look upon, from morning till evening, let me go where I will, will still keep gliding by me in unnumbered guises, and soothe me whilst I am busy at other things, and am thinking nothing of her." But the sensuous pleasure afforded by light is enhanced, while a purer and higher enjoyment is superadded, when the real meaning of the display is understood. As the astronomer sees in the sun a more glorious object than the sun of the poet, recognising in imagination not only the visible splendour of that orb, but the mighty energy with which it is swaying the motions of a scheme of circling worlds, the wondrous activities at work throughout its entire frame, the inconceivable tumult which must prevail in that seemingly silent globe, so the glories of light, rightly understood, are far more impressive than as they appeal simply to the senses.

Consider, for instance, the real meaning of sunrise. The orb seemingly rising above the horizon, but, in reality, at rest, is the source of all the glory which is spreading over the fair face of earth. The atoms of that remote body, vibrating with intensest activity, send forth in all directions ethereal waves, and of these relatively but a very few, about one in two thousand millions, fall upon our earth, producing the phenomena of sunlight.

They have been little more than eight minutes on the road, but in that short time they have traversed more than 90,000,000 of miles. Were they to fall directly upon our earth, we should see few of the splendours which attend the uprising of the sun. The deep air clothing our earth receives the onward rushing waves, and reflects them in all directions. To use Biot's simile, "The air is a sort of brilliant veil, which multiplies and diversifies the sunlight by an infinity of repercussions." Nor is the wonder of the scene, or its effect in filling the mind with solemn and poetic thoughts, diminished—on the contrary, it is enhanced—by the recollection that the gradually growing glory of day is brought about by the slow turning of the mighty earth,—

> "that spinning sleeps
> On her soft axle, as she paces even,
> And bears us soft with the smooth air along."

But if this is true of a scene of terrestrial splendour, how much more fully may it be said of the glories of the heavens? No poet, if unaware of the real meaning of modern discoveries respecting the celestial bodies, can be moved by the starlit depths as the astronomer is, at least the astronomer whose study of science is not limited to mere observation and calculation. Hundreds of bright points of light sparkling, and sometimes varying strangely in colour, form, no doubt, a beautiful scene;

but the scene is not less beautiful, and certainly it is far more impressive, when we remember that every one of these points of light is a sun, mighty in attractive energy like ours, its whole surface glowing with fiery heat, and every particle of its substance constantly in motion, if not always in the fierce rush of cosmic hurricanes, yet with the ceaseless vibrations which generate the ethereal light-waves telling us of the star's existence.

There is one strange thought connected with the motion of light-waves through the ether of space which has not, I think, received the attention it deserves.

Every one knows that when we look at the heavens we do not see the celestial bodies where they are, but where they *were*, and again, not where they were at any one moment of time, but some where they were a short time ago, others where they were very long ago. But it is not so generally known, or remembered by those who do know it, that if light were not so active as it is the result would be that utterly incorrect pictures of the celestial depths would continually be presented to us. As matters actually are no orb in space can appear very far from its true place. We see the sun, for instance, at any moment, not where he is, but where he was (or rather towards the direction in which he lay) about eight minutes before. But as the real velocity of the earth, and therefore the apparent velocity of the sun, amounts only to about eighteen miles per second, the sun is only thrown about 9000 miles out of his true position, which is but about the ninetieth part of his diameter; so that we see the sun very nearly in his right place. Now it might seem that a star whose light takes, say, twenty years in reaching us, must be seen very far from its true place, supposing the star to be travelling along very quickly; and, in one sense, this is true. If such a star is moving at the rate of fifty miles per second, athwart the line of sight, it will be out of place by so considerable a distance as 315,000,000,000 of miles. Yet the star will appear very nearly in its true position, simply because, at the star's enormous distance from us, even the great distance just named is reduced to a very small apparent amount. Such a star would, in fact, be displaced by only about the thirtieth part of the sun's or moon's apparent diameters, or by about a fifteenth part of the distance separating the middle star of the Great Bear's tail from its small companion, sometimes called Jack by the Middle Horse. Thus the stellar heavens present very truly to us the positions of the stars; for such athwart motion as I have just imagined would be very much larger than the motion of far the greater number of the stars. But we only thus see the heavens truly pictured because of the enormous velocity with which light travels. If light swept along only at the rate of a hundred miles in a

Fig 2.—Sunset at Sea.

second (a velocity still far beyond our powers of conception), there would be no believing what we should see, for every star, and our own sun, and all the planets, and even our own companion planet, the moon, would be thrown in appearance very far from their true positions. If they were all shifted in position by the same amount and in the same direction the picture would still be true, in a sense, just as we see a true picture of an object at the bottom of a clear lake, though the picture is displaced by the refractive action of the water on the rays of light. But, in the imagined case, the sun, and moon, and planets, and stars would be shifted by different amounts and in different ways, simply because they are moving at different rates and in different directions. The scene presented to us would have been utterly untrue. Astronomy as a science could probably have had no existence in such a case. Assuredly it could have had no existence until students of the heavenly bodies had learned to accept as the first axiom of their science the doctrine that "Seeing is not believing."

A strange thought truly, that so active are the orbs peopling space, so swiftly do they rush onwards upon their orbits, that light, carrying its message at a rate exceeding six thousand times the velocity of the swiftest express train, would be utterly unable to give a true account of the position and movements of the celestial bodies. Fortunately light gives a true record, because the qualities of the cosmic ether are such that the message of light is transmitted hundreds of times more swiftly than the swiftest bodies in the universe travel onwards upon their orbits around each other or in space.

II.

SPACE

ALTHOUGH astronomy tells us in clearest words of the vast depths of space which surround our earth on all sides, we are not thereby enabled to realize their enormous extension. It is not merely that the unknown depths beyond the range of our most powerful telescopes are inconceivable, but that the parts of space which we can examine are on too large a scale for us to conceive their real dimensions. It is hardly going too far to say that our powers of actual conception are limited to the extent of space over which the eye *seems* to range in the daytime. Of course in the daytime, at least in clear weather, there is one direction in which the eyesight ranges over a distance of many millions of miles,—namely, where we see the sun. But the sense of sight is not cognisant of that enormous distance, and simply presents the sun to us as a bright disc in the sky, or perhaps rather nearer to us than the sky. Even the distance of the sky itself is under-estimated. A portion of the light we receive from the sky on a clear day comes from parts of the atmosphere distant more than thirty or forty miles from us; but the eye does not recognise the fact. The blue sky seems a little farther off than the clouds, but not much; the light clouds of summer seem a little but not much farther off than the heavier clouds of a winter sky; a cloud-covered winter sky seems a little farther off than heavy rain-clouds. The actual varieties of distance among clouds of various kinds are not much more clearly discerned than the actual varieties of distance among the heavenly bodies. The estimate formed of the distance of a cloud-covered sky overhead probably amounts to little more than a mile, and it is very doubtful whether the mind presents the remotest depths of a blue sky overhead at more than two miles. Towards the horizon the distance seems greater, and probably on a cloudy day the sky near the horizon is unconsciously regarded as at a distance of about five miles, while blue sky near the horizon may be regarded as lying at a distance of six or seven miles, the arch of a blue sky seeming to be far more deeply curved than that of a cloud-covered sky.

It is to distances such as these that the mind unconsciously refers the celestial bodies. We know that the moon is about 2,000 miles in diameter, but the mind refuses to present her to us as other than a round disc much smaller than those other objects in sight which occupy a much larger portion of the field of vision. The sun cannot be conceived to exceed the moon enormously in size, seeing that he appears no larger; and all the multitude of stars are judged by the sight to be mere bright points of light in reality as they appear to be.

How, then, can we hope to appreciate the vastness of space whereof astronomy tells us? To the student of science attempting to conceive the immensities of whose existence he is assured, the same lesson might be taught in parable which the child of St. Augustine's vision taught the Numidian theologian. As reasonably might an infant hope to pour the waters of ocean into a hollow, scooped with his tiny fingers in the sand, as man to picture in his narrow mind the length and breadth and depth of the abysses of space in which our earth is lost.

Yet, as a picture of a great mansion may be so drawn on a small scrap of paper as to convey just ideas of its proportions, so may the great truths which astronomy has taught us about the depths of space be so presented that just conceptions may be formed of the proportions of at least those parts of the universe which lie within the range of scientific vision, though it would be hopeless to attempt to conceive their real dimensions.

Thus, when we learn that a globe as large as our earth, suspended beside the moon, would seem to have a diameter exceeding hers nearly four times, so that the globe would cover a space in the heavens about thirteen times as large as the moon covers, we form a just conception of the size of the moon as compared with the earth, though the mind cannot conceive such a body as the moon or the earth really is. When, in turn, we are told that if a globe as large as the earth, but glowing as brightly as the sun, were set beside the sun, it would look a mere point of light, we not only learn to picture rightly to ourselves how largely the sun exceeds the earth, but also how enormous must be the real distance of the sun.

Another step leads us to a standpoint whence we can form a correct estimate of the vast distance of the fixed stars; for we learn that so enormous is the distance of even the nearest fixed star, that the tremendous space separating the earth from that star sinks in turn into the merest point, insomuch that if a globe as bright as the sun had the earth's orbit as a close fitting girdle, then this glorious orb (with a diameter of some 184,000,000 of miles) would look very much smaller than such a globe as our earth would look at the sun's distance—would, in fact, occupy but about one-fortieth part of the space in the sky which she, though she would then look a mere point, would occupy if viewed from that distance.

But there is a way of viewing the immensities of space which, though not aiding us indeed to conceive them, enables the mind to picture their proportions better than any other. The dimensions of the earth's path around the sun sink into insignificance beside those of the outermost planets; but these in their turn dwindle into nothingness beside those of some among the comets. From the paths of these comets, if only sentient and reasoning beings could trace out in a comet's company those mighty orbits, and could have for the duration of their existence not the brief span of time which measures the longest human life, but many circuits of their comet home around the same ruling orb (as we live during many circuits of our globe around the sun), the dimensions of the star-depths, which even to scientific insight are all but immeasurable, would be directly discernible. Not only would the proportions of that mighty system be perceived, whose fruits and blossoms are suns and worlds, but even the gradually changing arrangement of its parts could be discerned.

Some comets, indeed, as I pointed out in an essay on comets several months ago (see Expanse of Heaven, p. 149), do not travel around the sun, but flit from sun to sun on journeys lasting millions of years, paying each sun but a single visit. A being inhabiting such a comet, and having these interstellar journeys as the years of his existence, so that he could live through many of them, would have a wonderful insight into the economy of the stellar system. If his powers of conception as far exceeded ours as the range of his travels and the duration of his existence, he would be able to recognise the proportions of a large part of the stellar universe as clearly as we recognise the proportions of the solar system.

But leaving these wonderful wanderers, whose journeys are as far beyond our powers of conception as the immensity of the regions of star-strewn space, we may find, among the comets belonging to the sun's domain, bodies whose range of travel would give their inhabitants far clearer views of the architecture of the heavens than even the profoundest terrestrial astronomer can possibly obtain.

Such a comet as Halley's (fig. 3) for instance, though one of comparatively limited range in space, yet travels so far from the sun that, from the extreme part of its path, it sees the stars displaced nearly twenty times as much (owing to its own change of position) as they are from the earth on opposite sides of her comparatively narrow orbit. And the length of this comet's year, if it indicated the length of the lives of all creatures travelling along with it, would suggest a power of patiently watching the progress of changes lasting not a few of our years only, but for centuries. Seventy-five or seventy-six years elapse between each return of this comet to the sun's neighbourhood, and one who should have lived during sixty or seventy circuits of this body around its mighty orbit would have been able to watch the rush of stars, with their velocities of many miles per second, until visible displacements had taken place in their positions.

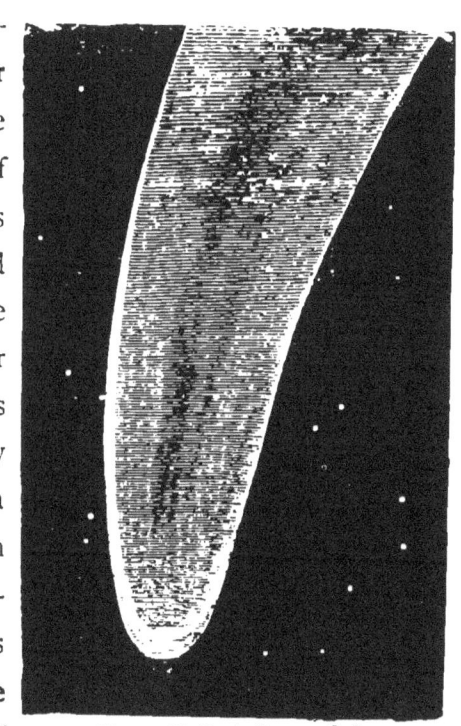

Fig. 3—Halley's Comet of 1835.

This, however, is as nothing compared with the mighty range in space and the enormous period of the orbit of the great comet of the year 1811 (fig. 4). This comet is, on the whole, the most remarkable ever known. It was visible for nearly seventeen months, and though it did not approach the sun within 100,000,000 miles, and was therefore not subject to that violence of action which has caused enormous tails to be thrown out from comets which have come within a few million miles of him, or even within less than a quarter of his own diameter, it flourished forth a tail 120,000,000 of miles in length. Its orbit has, according to the calculations of the astronomer Argelander, a space exceeding the earth's distance from the sun 211 times, and thus surpassing even the mighty distance of Neptune fully seven times. It occupies in circuiting this mighty path no less than 3065 of our years (with a possible error either way of about forty-three years). So that, according to Bible chronology, this comet's last appearance probably occurred during the rule of the judge Tola, son of Puah, son of Dodo, over the children of Israel, though it may have occurred during the rule of his predecessor Abimelech, or during that of his successor Jair.* During one half of the enormous interval between that time and 1811 the comet was rushing outwards into space,

---

* It might be suggested that the appearance of this blazing comet among the stars drove the more superstitious of the Israelites at that time to the worship of star-gods, as we read how, during the judgship of Jair, they "served Baalim, and Ashtaroth, and the gods of Syria, and the gods of Moab, and the gods of the Philistines, and forsook the Lord and served not Him." To a people like the Jews, who seem to have been in continual danger of returning to the Sabaistic worship of their Chaldean ancestors, the appearance of a blazing comet may have been a frequent occasion of backsliding.

Fig. 4.—Comet of 1811.

reaching the remotest part of its path somewhere about the year 278 (A.D.), and from that time to 1811 it was on its return journey. It is strange to think, however, that though the remotest part of its path lay 211 times farther from the sun than the earth's orbit, yet even this mighty path, requiring more than 3000 years for a single circuit, cannot be said to have carried the comet into the star-depths. If the earth were to shift its position by the same enormous amount the nearest fixed star would have its apparent position changed only by about an eighth part of the apparent diameter of the sun or moon, or by about one-quarter of the distance separating the middle star of the Bear's tail from its close companion.

But this fact of itself is most strikingly suggestive of the vast distance of the stars. For consider what it means. Imagine the middle star of the Bear's tail to be the really nearest of all the stars instead of lying probably twenty or thirty times farther away. Conceive a comet belonging to that sun after making its nearest approach to it to travel away upon an orbit requiring 3000 years for each circuit. *Then* (supposing that star equal to our sun in mass), the comet, though rushing away from its sun with inconceivable velocity during 1500 years, would, at the end of that vast period, seem to be no farther away than one-fourth of the distance separating the sun from its near companion. Look at the middle star of the Bear's tail on any clear night, and on its small satellite, remembering this fact, and the awful immensity of the star depths are strongly impressed upon the mind. But the observer must not fail to remember that the star really is many times more remote than we have here for a

Fig. 5—Six-tailed Comet of 1744.

moment supposed, and that such a comet's range of travel would be proportionately reduced. Moreover, many among the stars are, doubtless, hundreds, even thousands, of times still farther away.

Let us turn lastly to the amazing comet of the year

1744, pictured, at the time, as shown in fig. 5 (though probably the drawing is greatly exaggerated). We find that though it had the longest period of any which has ever been assigned to a comet as the result of actual mathematical calculation, yet its range in space would scarcely suffice to change the position of the stars in such sort that the aspect of the familiar constellations would be materially altered. Euler, the eminent mathematician, calculated for this comet a period of 122,683 years, which would correspond, I find, to a distance of recession equal to 2469 times the distance of the earth from the sun, or about eighty times the distance of Neptune. Yet this is but little more than twelve times the greatest distance of the comet of 1811. Probably the actual range of such an orbit from the middle star of the Bear's tail would be equal in appearance to the range described above on the supposition that the star is no farther from us than the nearest known star (Alpha Centauri). That is, such a comet, if it could be seen and watched during a period of about 122,000 years, would seem to recede from the star to a distance equal to about one-fourth the space separating it from its close companion, and then to return to the point of nearest approach to its ruling sun.

Such are the immensities of star-strewn space! The journey of a comet receding from the sun with inconceivable velocity during hundreds of thousands of years carries it but so small a distance from him compared with the distance of the nearest star as scarcely to change the appearance of the celestial landscape; and yet the distances separating the sun from the nearest of his fellow suns are but as hair-breadths to leagues when compared with the proportions of the scheme of suns to which he belongs. These distances, though so mighty that by comparison with them the inconceivable dimensions of our own earth sink into utter nothingness, do not bring us even to the threshold of the outermost court of that region of space to which the scrutiny of our telescopes extends. Yet the whole of that region is but an atom in the infinity of space.

III.

OF THE INFINITELY MINUTE.

WHEN I speak of the infinitely minute, I use the word infinitely not in its absolute sense, but relatively. Actual infinity of minuteness is as utterly beyond our conceptions as actual infinity of vastness. But we may speak of what is very much less than the least object of which our senses can make us directly conscious as *for us* infinitely minute. Among the greatest wonders science has to deal with are those relating to bodies and movements thus beyond the direct ken of our senses. There is a universe within the universe which our senses reveal to us,—a universe whose structure is so fine that the minutest particle which the microscope can reveal to us is, by comparison, like one of the suns which people our universe compared with the unseen particles constituting matter.

It is a strange thought that the objects constituting

our universe, so long regarded by man as the only universe, are in a sense pervaded by the materials of an utterly different universe,—which yet is as essential to our very existence as what we commonly call matter. We cannot live without light and heat, for instance, and again, light and heat affect matter as we know it; but they thus exist and affect such matter by means only of a form of matter unlike any which we can conceive. It is certain that if absolute vacancy separated our earth from the sun, even by the narrowest imaginable gap, his heat and light could never reach us. They could on more pass that vacant space than the wave-motion of water can cross a space where water itself is wanting. It is because of relations such as these that it has been said, and justly, that matter is the less important half of the material constituting the physical universe.

Our knowledge of this universe within our universe has been obtained within comparatively recent years. Men were unwilling or at least they spoke and thought as if they were unwilling, to believe that the universe of matter which they had so long recognised was dependent on another universe for its chief if not all its properties. They regarded heat as some sort of substance, which might, with more delicate means than they possessed, admit of being dealt with as chemists had dealt with the gases. The sun was full of this fluid, this phlogiston, as it was called. Light, in so far as it could be distinguished from heat, was another fluid; electricity was another. These were the imponderables, or unweighable substances of last century's science,—not as with us, the effects of modes of motion taking place in a universe which, though material, is yet not made of matter such as we know, or even such as we can at present conceive.

This is the greatest of all human scientific marvels,— the greatest because it includes all others. We know of a universe which is as infinite in extent, and doubtless in duration, as our own universe; which pervades all forms of matter: and yet we know of this universe only indirectly; by the effects of movements taking place within it, not by any perception of these movements themselves. Waves are ever beating upon the shores of our material universe, and constantly changing the form and condition of the coast line, but the waves themselves are unseen. We only know of their existence through the changes wrought by them.

We speak of the ether of space, and of waves traversing it, as though the ether were simply some fluid very much more attenuated than the rarest gas, even in a so-called vacuum. But in reality, so soon as we attempt to apply to the movements taking place in such an ether the mechanical considerations which suffice for the motions of all ordinary forms of matter, we perceive that it must of necessity be utterly unlike any kind of substance known to us. For instance, we find that though it is like a gas in being elastic, its elasticity is infinite compared with that of any material gas. Again, it is like a solid in retaining each of its particles always very near to a fixed position; but again, no solid we know of can be compared with it for a moment as respects this kind of rigidity. It is at once infinitely elastic and infinitely rigid. We cannot, for example, explain the phenomena of light unless we suppose the elasticity of the ether at least 800,000,000,000 times greater than the elasticity of air at the sea-level; and yet, as Sir J. Herschel long since pointed out, every phenomenon of light points strongly to the conclusion that none of the particles of the ether can be "supposed capable of interchanging places, or of bodily transfer to any measurable distance from their own special and assigned localities in the universe. Again, how are we to explain the continuance of the ether in its present condition, when we recognise the fact that a gas of similar elastic power would expand in all directions with irresistible force, diminishing correspondingly in density; yet the ether of space remains always, so far as we can judge, absolutely unchanged in position. Its characteristics certainly remained unchanged. Light travels at the same rate now as it did last year, last century, a million years ago. The ether, then, that bears it has presumably remained unchanged. If it were gaseous, and bounded on all sides by vacuum, it would expand with inconceivable velocity. To suppose it infinite in extent is to get rid of the difficulty perfectly; but only by introducing a difficulty far greater."*

A wonderful feature of the infinitely tenuous ether is, that while its ultimate particles must be inconceivably more minute than the ultimate atoms of ordinary matter, the movements taking place in it are transmitted with enormous velocities. The structure of our universe is on a grander scale; its least atom may comprise millions of millions of the largest component

---

* I do not say we can in any way avoid this far greater difficulty. Our own material universe cannot even be conceived as limited in any way save by void space of infinite extent; and it is as impossible for us to conceive an infinite void as to conceive the infinite extension of matter. Some modern mathematicians, indeed, assert that space is not necessarily infinite, but they accompany the assertion (very justly) with the admission that we cannot possibly conceive any boundary to space; and as one of the things they ask mathematicians to admit is the possibility that a straight line indefinitely produced both ways will at length re-enter into itself, while another is the possibility that in other parts of the universe two and two may make three or five, they are not likely, I conceive, to persuade most mathematicians (profoundly mathematical though they are themselves) that the mystery of infinity has been as yet entirely expounded.

portions of that infinitely tenuous ether. But amid that ether motions are transmitted with velocities transcending all but infinitely those which take place among the particles of matter composing the universe in which we "live and move and have our being." The planets, immense aggregates of matter such as we know it, sweep onwards upon their immense orbits, traversing many thousands of miles in an hour; but light and heat sweep along the ether of space, and by virtue of motions taking place within that ether at the rate of many tens of thousands of miles per second. The suns which people space rush onwards with mightier momentum, but less swiftly than the planets in their orbits. Comets attain the greatest velocities of all the bodies that science deals with, rushing sometimes, in their periastral swoop, with a velocity of hundreds of miles per second,—though yet in mid-space the comets of widest orbital range lag slowly enough, insomuch that some of those which, when nearest our sun, travel at the rate of two or three hundred miles per second, move more slowly when very far from him than many of our rivers. Taking even the swiftest rush of a comet within the solar domain, we find that light speeds along five hundred times more quickly,—so that if we represent the velocity of light by that of an express train (reducing light's velocity in scale to about one-10,000,000th part of its real value), the velocity of the most swiftly-moving comet would be represented by that of a walk at the rate of one-eighth of a mile per hour,—a very slow walk indeed.

It is not only amid the depths of space that these wonderfully swift motions take place in the ethereal universe. As I have said, that universe pervades ours throughout its entire extent. The densest of our solids is as freely traversed by the ether as a forest by the summer breeze. As the foliage of a thick forest may prevent the passage of fierce winds, so may a solid body prevent the passage of light-waves—though all solid bodies, as we know, do not prevent, and some scarcely even modify, the passage of light. But substances which prevent the passage of light are yet found capable of transmitting ethereal motions of similar velocity. According to Wheatstone's experiments electricity travels at the rate of more than 200,000 miles per second along stout copper wire. Fizeau's experiments gave a lower speed; but they did not negative Wheatstone's, the conditions not being the same. Can anything be more wonderful than the thought of the transmission of electricity with this enormous velocity? What really happens we do not know. Perhaps if we were told what really takes place between and among the particles of the wire, we should find ourselves utterly unable to conceive it—for, as we have seen, the properties of the ether, and, therefore, the processes taking place in the ethereal universe, are probably unlike any within our experience. But this we know—a certain condition of the molecules of the wire is transmitted, by virtue of the ethereal medium pervading the wire, at a rate so enormous that, if the wire itself could move at that rate, the force required to bring its mass to rest would suffice to generate enough heat to turn many times as much metal into the vaporous state.

Nay, even as regards the energy of their action on the matter of our universe, these movements in the ethereal universe enormously exceed the forces we are accustomed to regard as most powerful. The effects produced by gravity, for instance, are almost evanescent compared with those produced by heat. The sun's rays poured on a piece of metal for a few minutes produce motions in every one of the ultimate particles of the metal. Each particle vibrates with inconceivable rapidity (referring to the rate at which the vibrations succeed each other), and with great actual velocity of motion. Summing up the energy thus pervading the piece of metal, we find that it incalculably exceeds the energy represented by the velocity which the sun's attraction would communicate in the same interval to that piece of metal, supposed to be entirely under its influence at the earth's distance from the sun.

Or take another instance. "Think for a moment," say the authors of the "Unseen Universe," "of the fundamental experiments in electricity and magnetism, known to men for far more than 2000 years,—the lifting of light bodies in general by rubbed amber and of iron filings by a loadstone. To produce the same effect by gravitation-attraction,—at least, if the attracting body had the moderate dimensions of a hand-specimen of amber or loadstone,—we should require it to be of so dense a material as to weigh, at the very least, 1,000,000,000 pounds, instead of (as usual) a mere fraction of a pound. Hence it is at once obvious that the imposing nature of the force of gravity, as usually compared with other attractive forces, is due, not to its superior qualitative magnitude, but to the enormous masses of the bodies which exercise it."

We may put this illustration in another form. When we place a powerful magnet near a piece of iron, say at a distance of one inch, and the magnet lifts that piece of iron by virtue of its attractive power, a contest has been waged, if one may so speak, between the attractive powers of the small magnet and of the mighty earth, and the magnet has conquered the earth. Now the magnet has been much nearer than the earth to the piece of iron, for we know that the earth's attractive influence has been the same as though the entire mass of the earth were gathered at its centre, say 4000 miles from the piece of iron. A distance of 4000 miles contains 4000 times 1760 times thirty-six inches, or, roughly, 250 millions of inches. (This is in truth very near the true number of inches in the earth's radius, insomuch that many suppose the inch to have been originally taken as the 500,000,000th part of the earth's diameter. A British inch is about one-500,000,000th part of the polar diameter of the earth.) Since attraction diminishes as the square of the distances increases, and *vice versâ*, it follows that if the earth's entire mass could act on the piece of iron, at a distance of one inch, the attraction would exceed that actually exerted by the earth 250 million times 250 million times, or 62,500 millions of millions of times. In this degree, then, the earth is at a disadvantage compared with the magnet as respects distance. And one-62,500,000,000,000,000th part of the earth's mass would be capable of attracting the piece of iron as strongly as the earth actually attracts it, if that fraction of the earth's mass could exert its pull from a distance of only one inch. But a 62,500,000,000,000,000th part of the earth would be an enormous mass. It would weigh about 97,500 tons.

or some 218 millions of pounds. Thus a magnet which a child can lift exerts a greater attraction on the piece of iron at the same distance than a mass at least 1000 million times its weight could exert by its gravity only.

In fact we see from this illustration that gravity, though it produces effects so tremendous, though it sways the moon round the earth, the earth and all the other planets around the sun, and urges the sun and his fellow-suns through space, is, after all, but a puny force in itself. A child can lift his own weight against the attraction of the mighty earth; and by combined strength as many children as would have a weight equal to the earth's would easily bear a weight exceeding the earth's, if the force could be wholly and directly applied to such work.*

The attraction of gravity must, however, be regarded

---

* Of course the reader will understand that when I here speak of the earth's weight, I mean simply the pressure which would be exerted by the quantity of matter contained in the earth, if each portion were only subjected to an attractive force equal to that of gravity at the earth's surface. The actual force with which the earth is drawn in any direction, as a weight at the earth's surface is drawn downwards, depends on the distance and mass of the attracting body as well as on the mass of the earth; and strictly speaking, we ought not to say that the earth weighs so many millions of tons, but that she contains so many million times as much matter as a mass which at her surface weighs a ton.

as only one manifestation of the energies of the infinitely minute. It is in this sense well worthy of careful study. I propose to present in a future paper some of the strange thoughts which are suggested by the action of this wonderful force, the range of whose activity is seemingly co-extensive with the material universe.

IV.

THE MYSTERY OF GRAVITY.

THE law of gravity, or of the mutual attraction of masses of matter upon each other, accounts so perfectly for all the observed motions of the heavenly bodies, that we are apt to regard Newton's discovery of the great law as though it had finally solved the mystery of these motions. Many accept the verdict given by the poet Pope in the famous epitaph which he suggested for Newton,—

> "Nature and Nature's laws lay hid in night:
> God said, *Let Newton be!* and all was Light."

But Newton, who probably knew as much about his work as Pope, was of another opinion. Every one knows how he compared himself to a child who had picked up a few shells on the shore, while the ocean of truth lay unexplored before him. He has, however, spoken definitely of the great discovery which has rendered his name illustrious, in terms which show that *he* did not find that all was light. Among the questions which he specially would have had answered, amongst the secrets of nature concealed beneath the ocean of truth, the mystery of gravity was probably the chief. When Newton asked of the Ocean of Truth what Mrs. Hemans later said, and in another sense, of the natural sea—

> "What hidest thou in thy treasure-caves and cells,
> Thou hollow-sounding and mysterious main?"

he had in his thoughts the very power which he is commonly supposed to have explained, but which was in truth for him, more than for any man that had ever lived, the mystery of mysteries.

It may be well to consider the very words of the great philosopher, so far at least as our more diffuse language can present the concise expressions of the original Latin:

"Hitherto we have explained," he says, "the phenomena of the heavens and of our sea by the power of gravity, but have not yet assigned the cause of this power. This is certain" (we must hearken attentively here, for when a man like Newton speaks of aught as certain, we have sure ground to go upon),—"this is certain, that it must proceed from a cause that penetrates to the very centres of the sun and planets, without suffering the least diminution of its forces; that operates, not according to the quantity of surfaces of particles on which it acts (as mechanical causes usually do), but according to the quantity of the solid matter which they contain, and propagates its virtue on all sides to immense distances, decreasing always as the squares of the distances. Gravitation towards the sun is made up of the gravitations towards the several particles of which the body of the sun is composed, and in receding from the sun decreases accurately as the square of the distances as far as the path of Saturn . . . . , nay, and even to the remotest parts of the paths of comets . . . . But hitherto I have not been able to discover the cause of those properties of gravity from phenomena; and I frame no hypotheses:* for, whatever is not deduced from phenomena is to be called an hypothesis; and

---

* The words of Newton, "Hypotheses non fingo," have been often quoted in such sort as to give an entirely incorrect idea of his real opinion as to the relation between theoretical and practical science. As too commonly understood, they would, in fact, make his discovery of gravitation a great exception to his own rule. They must be taken in connection with his definition of a hypothesis, as "whatsoever is not deduced from phenomena." It is a part of true science, nay, it is the highest office of the student of science to deduce theories from phenomena. Such research stands as high above the simple observation of phenomena as architecture stands above brick-making or stone-cutting. But to frame hypotheses as the old Greeks did, trusting to the power of the understanding independently of the observation of phenomena, is to make bricks without straw and to build with them upon the sand.

hypotheses, whether metaphysical or physical, whether of occult qualities or mechanical, have no place in experimental philosophy. . . . To us it is enough that gravity does really exist, and act according to the laws which we have explained, and abundantly serves to account for all the motions of the celestial bodies and of our sea."

"Hitherto I have not been able to discover the cause of the properties of gravity." Such is the simple statement of the man who discovered those properties.

And now let us inquire a little into this law of gravity, not with the hope of explaining this great mystery of nature,—though, for my own part, I believe that the time is not far distant when the progress of discovery will enable man to make this approach towards the mystery of mysteries,—but in order to recognise the real nature of the mystery, which is a very different thing from explaining it.

In the first place the study of gravity brings us at once to the consideration of the infinitely minute,—at least of what is for us practically infinite in its minuteness. If we consider the above quotation attentively, we perceive that this quality of gravity was recognised by Newton. "It is not the quantity of the surfaces of *particles*," he says, "but the quantity of solid matter which they contain," that gives to gravity its power. Gravity resides in the ultimate particles of matter. We cannot conceive of matter so divided, no matter how finely, that non-gravitating particles could be separated from gravitating particles. Without entering into the question what atoms are, we perceive that these ultimate constituents of matter must contain, each according to the quantity of matter in it, the gravitating energy. Only, observe how incongruously we are compelled to speak. (It is always so when we deal with the infinite, whether the infinitely great or the infinitely minute.) We are speaking of atoms as the ultimate constituents of matter, and yet we are compelled, in describing their gravitating energy, to speak of the quantity of matter contained in each atom, —in other words, we speak in the same breath of an atom as not admitting of being divided or diminished, and of its containing matter by quantity, that is, by more or less. May we not, however, reasonably accept both views? The reasoning is sound by which science has proved that, so far as our material universe is concerned, there is a limit beyond which the division of matter cannot be supposed to go,—insomuch that Sir W. Thomson has indicated the actual limits of size of the atoms composing matter. Yet, passing in imagination beyond the bounds of our visible universe, and so entering into the next order of universe below it (in scale of construction), —the ether of space,—the atoms of our universe may be infinitely divisible in that universe, may be, in fact, compared with *its* particles, as the suns and worlds of our universe are to our atoms and molecules.

But while gravity thus draws us to the contemplation of the infinitely minute, it also leads us to the consideration of what is for us the infinitely vast.

Newton was only able to speak confidently of the action of gravity at the distance of Saturn, the remotest planet known in his day. He did, indeed, refer to the comets as probably obeying, even in the remotest parts of their paths, the force of the sun's gravity; but he could not be certain on that point, because in his time no comet had been proved to travel back to the sun after receding to the remotest portion of its track. We now know not only that the sun's attraction extends to the farthest parts of the solar system, having thus a domain in space nearly thirty times larger than the sphere of Saturn, but we perceive that many among the stars exert a similar force; for around them travel other stars even as the planets travel around the sun. Thus we know that gravity is exerted in regions lying hundreds of thousands of times farther from the sun than Saturn is. We have, indeed, every reason to believe, not only that star unto star extendeth this mysterious attractive influence, but that the least particle in the inmost depths of sun or world exerts in full force on each particle, even of suns lying millions of times beyond the range of the most powerful telescope yet constructed by man, the full energy corresponding (i.) to the quantity of matter in itself and such particle, and (ii.) to the distance separating each from each.

This is amazing enough; but there is something more perplexing and mysterious in gravity even than this. Not only does gravity lead us to consider the infinitely minute in space on the one hand, and the infinitely vast in space on the other, but also it leads us to consider the infinitely minute and the infinitely vast in time also, and this in such a way as to suggest a difficulty which, as yet, no man has been able to solve.

Light travels, as we know, with a velocity so enormous, that, by comparison with it, all the velocities we are familiar with seem absolutely as rest. But gravity acts so quickly that even the velocity of light becomes as rest by comparison with the velocity of the propagation of gravity. Laplace had occasion, now nearly a century ago, to inquire whether a certain change in the moon's motion, by which she seemed to be gradually hastening her motion round the earth, might not be caused by the circumstance that gravity requires time for its action to be propagated over great distances. He found that if the whole of that change had to be explained in this way, which would be giving to gravity the slowest admissible rate of transmission, the velocity with which gravity is propagated would be *eight million* times greater than the velocity of light. If, on the other hand, that change in the moon's motion could be satisfactorily explained in some other way, then the velocity of gravity must be at least 16,000,000 times greater than the velocity of light. He himself soon after discovered what was in his day regarded as a complete explanation of the hastening of the moon's motion; and though in our own time Adams of Cambridge has shewn that only half the hastening can be accounted for by Laplace's reasoning, the general explanation of the remaining half is that it is not a real hastening of the moon, but is only an apparent hastening caused by the gradual slowing of the earth's rate of turning on her axis. This makes the *day* by which we measure the moon's motion seem longer (very slightly, however).* Supposing, however, half the moon's hastening were left unexplained, and that the non-instantaneous transmission of gravity were the only way of accounting for it, even then it would be certain that gravity is propagated at a rate exceeding 12,000,000 times the velocity of light.

Indeed, at present, owing to the more exact observa-

---

* The point is explained in a paper called "Our Chief Timepiece Losing Time," in the first series of my "Light Science for Leisure Hours."

tions available, and the greater range of time over which they extend, it may safely be said that the rate of propagation of gravity is far greater than this. It is even held by some that gravity acts instantaneously over any distance, however vast.

Although I cannot here indicate the exact nature of the reasoning by which the enormous rapidity of the action of gravity is inferred, I must briefly indicate the general argument, that the reader may not suppose the matter to be merely speculative. Suppose that the action of gravity were propagated at the same rate as light.

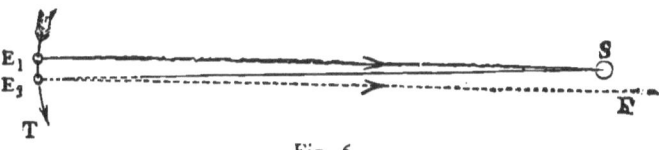

Fig. 6.

Then the earth would feel the pull of the sun eight minutes or so after she had been in the place where the sun began to exert that particular pull. The direction of the pull then would not be that of the straight line connecting the earth and sun at the moment when the pull was felt, but that of the straight line connecting the sun and the earth eight minutes or so before. For instance, when the earth is at E_1, fig. 6, the sun at S would begin to exert a pull in the line E_1 S, but the earth would only feel this pull when she got to E_2, her place eight minutes later, when it would act upon her in the direction E_2 F, parallel to E_1 S. Now this pull, E_2 F, may be divided into two parts, one along E_2 S, pulling the earth towards the sun S, the other along E_2 T in the earth's course, *hastening her therefore*. But the maintenance by the earth of the same constant track depends entirely on the action of gravity sunwards. If there is any action in addition, hastening the earth, then she will not keep her course,* but will travel in a constantly widening path, —or, in a sort of spiral, very slowly retreating from the sun, but retreating constantly. The change of distance would not be measurable in millions of years; but the increase in the length of the year *would*, before long, be observable. Because there is no such increase, astronomers feel well assured that gravity is not only propagated more swiftly than light, but many times, even, as we have seen, many millions of times, more swiftly.

It is then in an infinitely minute time that the action of gravity traverses all ordinary distances. The earth's

---

* In the popular, but incorrect way of speaking, the balance between the centrifugal and the centripetal force will no longer be maintained: the increase of velocity will give the centrifugal force the advantage, and it will slowly draw the body away from the centre. In reality there is no centrifugal *force*, the only force acting on the earth in her course round the sun being the sun's attraction upon her, which, however, must keep bending her course from the straight line, if she is to maintain her distance. In the case above imagined it would not bend her course actively enough.

pull on the moon takes less than the 50,000,000th part of a second in reaching the moon,—and the particles constituting the mass of the earth act on ourselves, and on all the objects which lie near the earth's surface, in far less than the 10,000,000th part even of this utterly minute time-interval.

Yet age after age has passed during which this infinitely active force has been at work without diminution, and age after age will continue to pass without any change in its activity. For millions of millions of æons it has lasted and will last, so permanent is it; while its operation is felt simultaneously at points millions of millions of star-distances apart. What infinities of distance has this wonderful attractive force traversed!

But even these considerations do not present the greatest of the marvels of gravity. It is wonderful, indeed, to consider a form of attraction possessed by the infinitely minute, and exerted over the infinitely vast, operating in portions of time immeasurably small, and extending its operations throughout time infinite. But the mystery of mysteries is not here. The marvel of marvels is this, that, so far as we can perceive, the force of gravity is exerted without any material connection with the objects moved by it. Matter seems to act where it is not, to use the phraseology of the schools.

Of this "action at a distance," Newton himself said, that it is inconceivable, that in point of fact it is impossible. "No man," he said, "who has, in philosophical matters, a competent faculty of thinking," can "for a moment believe that a body can act through a vacuum, without the intervention of anything else by or through which the force may be conveyed from one body to another." Yet this is precisely what gravity seems to do. The ether occupies, indeed, all space; but there is nothing at present known to us by which we can understand how the either can transmit the force of gravity. The power of the ether in the rapid transmission of undulations seems to attain its limit in the propagation of light and heat and electricity at the rate of nearly 200,000 miles per second. How the ether can act so as to serve as a medium of communication between bodies at all distances, transmitting impressions 10,000,000 times faster, at least, than light travels, nothing at present known to us enables us to say. I have, in a lecture which I gave in America upon the mysteries of the universe, indicated a way in which gravity may be conceived to be generated and transmitted; and I may hereafter describe the conception (based partly on the views of Le Sage). But it is only a conception. There is no phenomenon (except the very form of attraction which has to be explained) tending to show that the conception is correct. And even if it be accepted, it brings us face to face with only greater marvels.

At present, however, let this simply be said in conclusion—that the apparent action of gravity at a distance is, of all physical wonders, the greatest yet known to man. If we accept the opinion of Newton, which, indeed, seems to me indisputable, that matter cannot act through a vacuum, then we must admit the existence of properties, as yet unthought of, in the ether of space, or in some still more subtle universe permeating that ether. If, on the other hand, we accept the belief that matter can act at a distance, then is there no miracle, either of those believed in by mankind generally, or of those more generally rejected, which exceeds in marvellousness this wonder of all the wonders of physical science.

V.

THE END OF MANY WORLDS.

 SIGN has recently appeared in the heavens which has been interpreted in a way suggesting that many worlds like our own have undergone a terrible catastrophe, every living creature upon them being consumed as by fire. I propose briefly to consider some of the thoughts suggested by this strange event.

It is difficult when we look at the star-lit heavens, suggestive as they are of solemn peace, to conceive the stupendous energy, the fierce uproar and tumult, of which even the faintest visible star in reality tells us. Pythagoras spoke of the harmony of the celestial spheres, which we are only prevented from hearing by its continuity. "There's not the smallest orb which thou beholdest," said the science of the middle ages,

"But in his motion like an angel sings,
Still quiring to the young-eyed cherubim."

The science of our own time tells us a still stranger story. There's not the smallest orb which thou beholdest, she says, but in his motion throbs like a mighty heart, still pulsating life to the worlds which circle round it. But while our powers of vision are limited to the narrow range of our present telescopes, we cannot watch the action of these great centres of energy, nor can we hope that the uproar of those remote fires will ever reach mortal ears, though to the mind's ear clear and distinct. It is no longer a mere fancy that each star is a sun. Science has made this an assured fact, which no astronomer thinks of doubting. We know that in certain general respects each star resembles our sun. Each is glowing like our sun with an intense heat. Around each, as around our sun, are the vapours of many elements. In each the fires are maintained, as they are maintained in our sun, in some way which may be partly mechanical, partly chemical, but which certainly does not in the least resemble combustion. We know that in each star processes resembling in violence those taking place in our own sun must be continually in progress, and that such processes must be accompanied by a noise and tumult compared with which all the forms of uproar known upon our earth are as absolute silence. The crash of the thunderbolt, the bellowing of the volcano, the awful groaning of the earthquake, the roar of the hurricane, the reverberating peals of loudest thunder, any of these, or all combined, are as nothing compared with the tumult raging over every square mile, every square yard, of the surface of each one among the stars.

If we remember this when we hear of stars varying in brightness, we shall perceive that the least change which could be recognised from our remote stand-point must represent an accession or falling off of energy corresponding to far more than all the energies existing on our earth, or indeed on all the members of the solar system taken together. Astronomers recognise our sun as in one sense a variable star; for we can hardly suppose that he shines with the same degree of brilliancy when many spots mark his surface as when he is quite free from spots; and astronomers know that these changes in the sun's condition correspond to wonderful changes in his activity. When spots are most numerous, the coloured flames rage with fierce energy over his whole globe, metallic vapours are shot forth from below his visible surface with velocities of many miles per second. Whereas, when he has no spots, the coloured flames sink down from their former height of tens of thousands of miles, till they are but a few thousand miles in height;

while metallic vapours are seldom emitted, and never to the same height, or with the same velocity, as when the spots are most numerous. But though the sun thus varies in condition, and probably in his total brightness, we cannot suppose that such variations could be recognised from the distance of even the nearest among the fixed stars. What, then, must be the nature of changes taking place in a star, that we, at our enormous distance, should be able to recognise them! We may well believe that the entire aspect of such a star must be changed to the inhabitants, if such there are, of worlds circling around them.

If, however, the changes taking place in stars, whose variations of brightness can just be recognised, must be amazing, how stupendous must be the changes affecting a star which alternates from brightness to invisibility, like Mira, the Star Wonderful, in the constellation of the Whale! how destructive those affecting a star like Eta, of the ship Argo, which has varied from the fourth magnitude to a lustre nearly equalling that of Sirius, and thence to the lowest limit of visibility, in the course of the last hundred years!

Even these changes, however, though justly regarded as among the chief wonders and mysteries of the stardepths, seem in turn to sink into nothingness by comparison with the sudden appearance of a new star, as interpreted by modern scientific observations. Of old, when a new star appeared, it was thought for awhile to be a fresh creation; a new sun set in the centre of a new system of worlds,—a thought which was not then so startling as in our own times it would be reckoned. When the new star was seen slowly to die out until at last it became invisible, men were content to regard it as a sign set in the heavens for a special purpose. Nor did they find much difficulty in associating such a phenomenon with some event of importance occurring during its continuance, or soon after the new star had died out. Such were the explanations offered respecting the exceedingly bright star which made its appearance in the constellation Cassiopeia in the year 1572. The place in which it appeared is shown in fig. 7. It must have sprung into its full glory in a very short time, for Tycho Brahé, the celebrated astronomer, tells us that, returning on November 1, 1572, from his laboratory to his dwelling-house, he saw the new star, which he was certain had not been visible an hour before, shining more brightly than any before seen. It surpassed all the stars in the heavens in brilliancy, and even Jupiter when that planet is at its brightest. Only Venus at her brightest was superior to the new star. For three weeks it shone with full lustre, after which it began slowly to decline. Being situated in a part of the heavens always above the horizon (for

European observatories), the star's entire history could be followed. It remained for sixteen months steadfast in its position like the other stars. As it decreased in size it varied in colour. "At first," says an old writer, "its light was white and extremely bright; it then be-

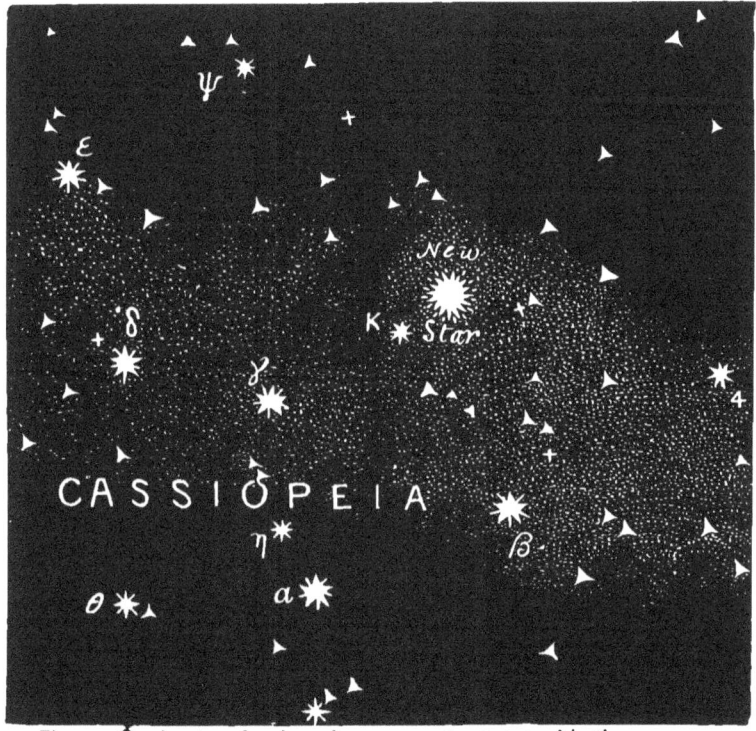

Fig. 7.—Cassiopeia; showing where a new star appeared in the year 1572.

came yellowish; afterwards of a ruddy colour like Mars; and finished with a pale, livid white, resembling the colour of Saturn."

In passing it may be remarked that there are reasons for expecting the return of Tycho Brahé's star in the course of a few years. For other new stars have been recorded as seen in the same part of the heavens in the years 945 and 1264, and though the interval from to 1264 (or 319 years) exceeds by 11 years the interval from 1264 to 1572 (or 308 years), yet the difference is but small by comparison with either entire interval; and we may not unreasonably believe that the three new stars seen in Cassiopeia have been only three apparitions of one and the same star, which shines out, with superior lustre, for a few months, once in a period averaging about 313 years. It seems to me not at all unlikely that, some time during the next twenty years, astronomers will have an opportunity of examining, with the telescope and spectroscope, a star which last appeared before either instrument had been invented.

Already facts are known respecting the so-called new stars which will not permit us to accept the explanations of old so readily offered and admitted, simply because so little was certainly known.

In the year 1866 a star appeared suddenly in the constellation of the Northern Crown, where no star had before been visible to the naked eye. It was a little below the arc of stars forming the celestial coronet.*

---

* Its place is indicated in my School Atlas, as well as (of course) in my Library Atlas, from the latter of which the small maps illustrating the present article have been pricked off. The new star is

It shone as a second magnitude star when first seen, but very rapidly diminished in lustre. It increased our knowledge in two important respects.

First, on examining Argelander's charts of the northern heavens, the new star was found to have been observed and charted as a tenth magnitude star, that is, four magnitudes below the lowest limit of naked eye vision. It was not, then, a new sun, though it might still truly be called a new star, in this sense, that it was a new member of the set of stars which adorn our skies as seen by ordinary vision.

In the second place, the star was subject to the searching scrutiny of spectroscopic analysis, with results of a most interesting character.

The reader is no doubt aware that when the light of a star is analysed into its component colours by the instrument called the spectroscope, it is found that all the colours of the rainbow are present, as in the case of solar light, but (also in the sun's case) not all the tints of these colours. Certain dark lines athwart the rainbow-tinted streak, called the *spectrum* of the star, indicate the

marked T in the Crown (Map VIII.), and must not be confounded with the star r, as in Roscoe's Treatise on Spectral Analysis, and in some astronomical works. The star r is a well known fifth magnitude star, which has shone with no perceptible increase or diminution of splendour since Bayer's time certainly, and probably for thousands of years before.

presence of absorbing vapours in the star's atmosphere. This general statement is true of every fixed star, though the dark lines of some stars differ in number and position from the dark lines of others, showing that other absorbing vapours are present. In the case of the new star in the Crown, the usual stellar spectrum was shewn,—a rainbow-tinted streak crossed by a number of dark lines. But besides these, there were seen four very bright lines, —lines so bright that the rainbow-tinted streak appeared as a dark background. The meaning of this is well understood by spectroscopists. It signifies that besides the vapours which, being cooler than the star, absorbed a portion of its light, and produced the dark lines, some vapours were present in the star's atmosphere which were a great deal hotter than the star, and so produced bright lines. Now two of the lines corresponded in position with two of the well known lines of the gas hydrogen, showing that this was one of the gases which had been raised to an unusual degree of heat.

It was inferred that there had been some tremendous disturbance in that remote star, by which the hydrogen and some other vapours present in its atmosphere had been intensely heated. But astronomers were unable to decide whether the disturbance was of the nature of a conflagration, the hydrogen actually burning, or whether the heat was occasioned in some other way, as by the downfall of some immense mass upon that remote sun. For burning hydrogen and glowing hydrogen, though either could give the observed bright lines, are very different things. In the former case a chemical change is taking place, as in the case of burning wood or coal; the latter case resembles that of redhot iron, which is not burning itself (not changing into a different form as everything does which burns), though it will burn other things,—in the ordinary, and incorrect, use of the expression.

The general belief was that there had been a downfall of matter on the star in the Crown, by which the whole globe of that sun had been excited to an intense degree of heat, especially at the surface, near which lies the hydrogen atmosphere of the star.

I must leave, however, to the next part, the further consideration of the strange thoughts suggested by the outburst of this star. I wish to use the small space remaining at present to indicate the place where another new star burst forth last November, so that any readers of these pages who have telescopes may know where to look for a sun which is now dying out, but was shining a few weeks ago as a third magnitude star. Fig. 8 presents a portion of the well-known constellation Cygnus or the Swan. Any star atlas will indicate the place of the lettered stars shown in the figure. The constellation itself does not show at all well at this season of the year.* The part shown in the figure is close to the horizon, and directly under the pole-star, at

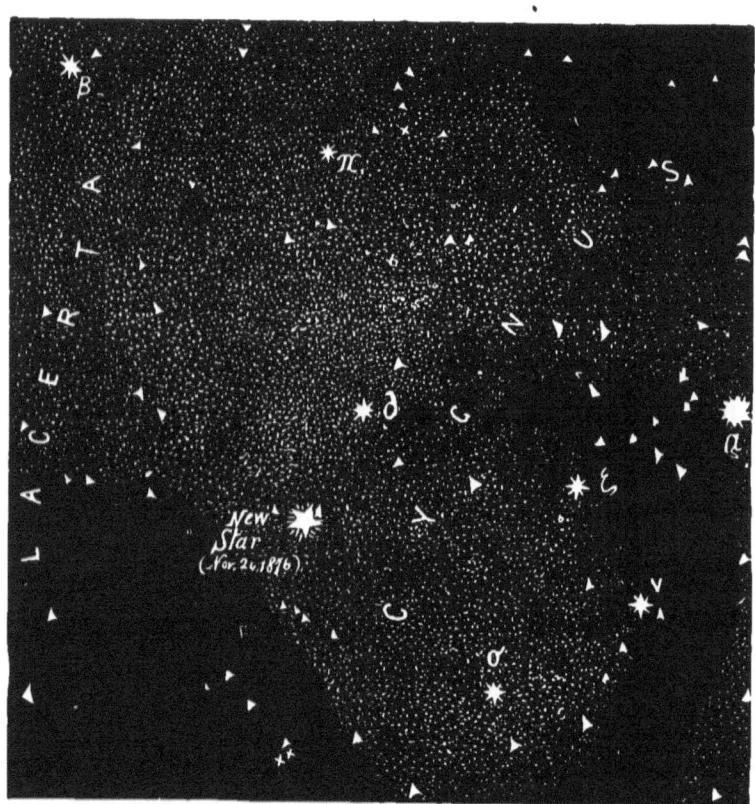

Fig. 8.—Part of Cygnus, showing the place of new star (November 24, 1876).

about half-past ten in the middle of February; but a little higher up, between north and north-east, at mid-

* This chapter was first published in February, 1877, when the star was already invisible to the naked eye.

night. Professor Schmidt, of the Athens Observatory, noticed a new star, in the place shown, on November 24th last. It must have shone out suddenly, for Schmidt had been observing in that region on the night of November 22nd (the last preceding clear night). It has since gradually faded, until now a small telescope is required to show it, shining as a seventh magnitude star, with a well-marked orange tint.

We have now to consider the history of this star, and discuss the general questions suggested by the sudden blazing out of suns which had for many years, and probably for many centuries, shone continuously with a far feebler lustre. It is clear that we have good reason to be interested in these questions, seeing that, for aught we know, our sun may be one of those exposed to sudden great increase of lustre.

It seems certain, in the first place, that this star leapt very suddenly to its full splendour. Schmidt had been observing the same regions of the heavens only two evenings before, and is sure the star was not then shining visibly to the naked eye. Again, astronomy is now studied by so many persons, and so many more who are not students of astronomy are now well acquainted with the constellations, that it is very difficult for a new star to shine many hours without being detected. For example, the new star in the Crown, which appeared in May, 1866, though not so well placed for observation, was detected by many observers at widely distant stations within a few hours of each other. It is probable that the star acquired its full lustre in a few hours at the utmost, and quite possible that, had any one been watching the place where the star appeared, he would have been able to see the star grow into full brightness by visible change of lustre, just as the lustre of a revolving light in a distant lighthouse visibly waxes and wanes. It may be, of course, that the increase of the star from its ordinary lustre, up to the stage when first it was visible to the naked eye, occupied many days, or even many months or years; but it seems more likely that as the later stages of increase were rapid, so also was the entire development of the new lustre. In that case, if there were inhabited worlds circling around that remote sun, they had but brief warning of the fate in store for them, as presently to be described.

Like the star in the northern Crown, the new star in Cygnus was subjected to the searching scrutiny of the spectroscope. The results, though similar in general respects, were even more interesting than in the case of the brighter new star. In the interval between 1866 and 1876 spectroscopic analysis has developed largely.

It has thus become possible to analyse more completely the light even of faint stars than the light of bright stars could be analysed a decade of years since.

The spectrum of the new star as examined by M. Cornu, of the Paris Observatory, showed the bright lines of hydrogen, indicating the presence of enormous quantities of glowing hydrogen, in a state of intense heat. But beside these bright lines, others also could be seen. One of these was an orange-yellow line. It will be understood that the faint spectrum of a star cannot be so readily lengthened by increasing the dispersion as a bright spectrum; for with too great dispersion the light fades out altogether. And though this is not strictly the case with the bright lines, which are merely thrown farther apart by dispersion, yet still it remains true that one cannot deal with a star spectrum even of bright lines as one can with the solar spectrum. So that M. Cornu was not able to determine whether the orange-yellow line belonged to sodium, or to that other substance, whatever it may be, which produces the orange-yellow line seen in the spectrum of a solar prominence.* Another bright line, green in

---

* It will be remembered by those familiar with the history of solar observation, that when the spectrum of the solar prominence was first observed, the orange-yellow bright line was supposed to be the well-known double sodium line. It is so near to this pair

colour, agreed in position with a triple line belonging to the metal magnesium. Lastly, a bright yellowish-green line was seen, which is known to be present in the spectrum of the sun's corona and of the low-lying ruddy matter round the sun, called the *sierra* by some, and by others (apparently unfamiliar with the Greek language) the chromosphere.

Now all this agrees very well with what had been noticed in the case of the star in the Northern Crown. For, unquestionably, if a sun increases so much in heat and lustre that the hydrogen outside it glows more brightly than the body of the star, then other matter *outside* that sun might also be expected to share the great increase of heat. We see that, outside our own sun, hydrogen, a certain unknown vapour of an orange yellow colour, magnesium, and another unknown vapour of greenish-yellow colour are present in enormous quantities; and it seems, therefore, reasonable to believe that other suns have these gases extending far outside the rest of their substance. It is certain that, if our sun were caused to glow with far more than its present degree of heat, the gases whose increase of brightness would be most discernible from a distant

---

of lines, that while they are called D 1 and D 2, it has been called D 3; and in a spectroscope of small dispersive power the three would be seen as one.

station (as a world circling around some remote star) would be just those gases which were glowing so resplendently around the star in Cygnus last November—or rather at the time when that light which reached us last November set out from the remote star in the Swan.

When we view the outburst of that remote sun in this way the thoughts suggested are not altogether satisfactory. That sun shows far too much resemblance to our own, and behaved, so far as can be judged, far too much as our own sun would behave if roused to many times its present degree of heat and splendour. When we hear of a railway accident it is a matter of special interest to us (if we travel much) to learn whether the conditions under which the accident took place resembled those under which the trains proceed by which we chiefly travel. When an express train suffers in such a way as to show some special danger arising from great velocity, we find ourselves to some degree concerned personally in the investigation which follows, if we travel generally by quick trains. If a bridge breaks down, and we have often to traverse bridges in railway journeying, we are similarly concerned, especially if any of the bridges we have to cross resemble in structure the one which has given way. So also of many other special forms of danger in railway travelling. Now, on the same principle, we cannot but regard with considerable interest the circumstance that, apparently, a catastrophe has taken place in the star in Cygnus, which has not only affected a sun resembling our own very closely in constitution, but has produced effects very closely corresponding to those which would affect our own sun if, through any cause, he were excited to many times his present degree of heat.

Let us pause a little to reflect upon the effects which would follow a great increase of the sun's lustre. A change in our own sun, such as affected the star in Cygnus, or that other star in the Northern Crown, would unquestionably destroy every living creature on the face of this earth; nor could any even escape which may exist on the other planets of the solar system. The star in the Northern Crown shone out with more than 800 times its former lustre: the star in Cygnus with from 500 to many thousand times its former lustre, according as we take the highest possible estimate of its brightness before the catastrophe, or consider that it *may* have been very much fainter. Now, if our sun were to increase tenfold in brightness, all the higher forms of animal life and nearly all vegetable life would inevitably be destroyed on this earth. A few stubborn animalcules might survive, and, possibly, a few of the lowest forms of vegetation, but naught else. If the sun increased a hundredfold in lustre his heat would doubtless sterilise the whole earth. The same would happen in other planets. The heat falling on the remotest members of the solar system would not, indeed, be excessive according to our conceptions. But if we regard Neptune, Uranus, Saturn, and Jupiter as the abode of life (which, for my own part, I consider altogether improbable), we cannot but suppose the orders of living creatures in each of these planets to be well fitted to exist under the conditions subsisting around them. If this is so—as who can for a moment doubt?—a sudden enormous increase in the sun's heat, though not making the supply received by those planets much greater than, or even equal to, the supply which we receive from the sun, would prove as fatal to living creatures there as to living creatures on our earth.

If, then, the sun increased in splendour as the stars have increased which the astronomers call new stars or temporary stars, there would be an end of life upon this earth; and nothing short of either the spontaneous development of life, or of the creation of various forms of life, could people our earth afresh. Science knows nothing of spontaneous generation, and believers in revelation reject the doctrine. Science knows nothing of the creation of living forms, but believers in revelation accept the doctrine. Certain it is that if our sun ever undergoes the baptism of fire which has affected some few among his brother suns, one or other of these processes (if creation can be called a process) must come into operation, or else our earth and her companion worlds would for ever after remain absolutely devoid of life.

But if our sun, without suffering so great a change, underwent a change of less degree, it might well happen that though there would be enormous destruction of life upon the earth and other planets, some life (presumably the strongest and best) would survive. In that case, after a long period of time, the earth would again be well peopled, and it might even be that the various races of terrestrial creatures would be improved, by the desolation which the great solar conflagration had wrought.

It is somewhat curious, considering how little there is in the ordinary progress of events to suggest the idea, that most of the ancient systems of cosmogony recognised the periodical destruction of living creatures on the earth by fire as well as by water. Each form of destruction was supposed to be brought about by planetary influences. The Ecpyrosis, or destruction by fire, was effected when all the planets were in conjunction with Cancer; the Cataclysm, or destruction by flood, when all the planets were in conjunction with Capricorn. Each form of destruction was supposed also to purify the human race. "Towards the termination of each era," writes Lyell, speaking of these old ideas, "the gods could no longer bear with the wickedness of men, and a shock of the elements or a deluge overwhelmed them; after which calamity Astrea again descended on the earth, to renew the golden age." The Greeks undoubtedly borrowed all such doctrines from the Egyptians, who "believed the world to be subject to occasional conflagrations and deluges, whereby the gods arrested the career of human wickedness, and purified the earth from guilt. After each regeneration mankind was in a state of virtue and happiness, from which they gradually degenerated again into vice and immorality."

Considering that we have every reason to believe the records of great floods to relate to events which actually occurred, however imperfectly remembered, it seems not unreasonable to believe that the tradition of great heats had its origin in observed phenomena. As neither ordinary conflagrations nor volcanic outbursts would have suggested traditions of the kind, it would seem not impossible that at certain times our sun may have acquired for a time unusual lustre and heat, causing great and widely spread destruction among all forms of animal and vegetable life.

This idea may possibly seem to many, especially at a first view, too wild to be entertained for a moment. Our sun shines, so far as appears to ordinary observation, with steadfast lustre from year to year, and also from age to age. If an occasional hot season suggests for a while to some that the sun has grown hotter, or a cool season that he has grown cooler, the restoration of cool or warmer weather, as the case may be, causes the thought to be quickly cast on one side that a change of either kind has taken place. Again, if we examine the historical records of past ages, we find little to suggest the idea, or even the possibility, that the sun in former times shone with greater splendour or with less than at present. The men of those days were formed like the men of our own day, and could not have supported any much greater degree of heat or of cold than men can support at present. Any sudden accession (or diminution) of solar light and heat, such as we are considering, would certainly have attracted marked attention, and have been recorded for the benefit of future ages. The geologic record, again, does, indeed, suggest variations in the sun's emission of heat as constituting one among the few available explanations of the existence of tropical forms of life in certain strata and of arctic forms in other strata. But even if this explanation be the true one, which is by no means established, such variations must of necessity have been slow, the condition of increased heat continuing for many ages in succession, and the like with the condition of diminished heat. We have no evidence, historical or geological, of the occurrence of any sudden accession of solar heat, followed by a quick return to the normal temperature, unless we find such evidence in the tradition prevalent among Egyptian, Indian, and Chinese cosmogonists, that at certain recurring epochs in the past our earth has undergone destruction and renovation by fire.

Yet, as I shall now show, it appears that the one only natural interpretation which can be given of the outburst of a new or temporary sun indicates an event which might happen to our own sun, and an event which if it happened at all would happen periodically. Moreover, while it will appear that there is no reason for fearing the possible occurrence (which would, in such case, be really the recurrence) of such a catastrophe in the case of our own sun as has affected the stars in the Crown and in Cygnus, there is no reason for rejecting as incredible the idea that catastrophes very serious in their character may have affected our sun; and there is abundant reason for believing that *small* alterations in the sun's total emission of light and heat take place very often, in some cases periodically; in others—so far as we can yet judge—periodically.

Lastly, it will be seen that there is always a possibility that our own or any other sun may undergo precisely such a change as the stars in Cygnus and the Northern Crown. Some indeed, even among men of science (as the Abbé Moigno, for example) believe that it was an event of this sort which St. Peter predicted when he wrote, that as the old world, being overflowed with water, perished, so "the heavens and the earth which are now, by the same word are kept in store, reserved unto fire." According to that view, the day of destruction will come "as a thief in the night; in the which the heavens shall pass away with a great noise, and the elements shall melt with fervent heat, the earth also, and the works that are therein shall be burned up."

Let us consider how the sudden brightness of a new star may be explained.

I must confess that for my own part I do not attach much weight to the suggestion once made by Mr. Huggins, that an actual conflagration had taken place in the case of the new star in the Northern Crown. It does not seem to me that any process of mere burning could account for the enormous accession of light and heat which that sun underwent.

Consider the case of our own sun. His heat is very far beyond that which would be given out by any matter known to us undergoing any known process of true combustion. That is to say, if a mass as large as the sun of any known substance were caused to burn, under any conditions we can imagine, the momentary emission of heat by that mass would be very much less than the momentary emission of heat by the sun.

Now it is quite conceivable that by some great accession of combustible matter, some supply of fuel exceeding many times his entire mass, the sun's entire emission of heat might be very largely increased. But though such an idea is conceivable, it seems altogether far-fetched. The conception is, in fact, inadmissible as an explanation of the increase of heat of a temporary star, not because of the improbability of the sudden accession of so enormous a quantity of matter (though that improbability is very great), but because if so enormous a quantity of matter fell upon the sun, many times as much heat would be generated by the mechanical effect of the impact as by the combustion of the freshly received matter. So that even with the daring assumption here made, combustion would account for only a small portion of the increase of light and heat.

Huggins' idea was indeed somewhat different. He supposed that in consequence of some great internal convulsion of the sun in the Northern Crown a large volume of hydrogen and other gases was evolved from the interior, the hydrogen then by burning giving out the light corresponding to the bright lines. At the same time, the mass of the sun would be intensely heated by the surrounding mass of glowing hydrogen. When the liberation of gas from the interior ceased the flame would die out, and the sun's surface would gradually cool. But if we judge by the case of our own sun, the heat of the *burning* hydrogen would be nothing near so great as the heat of the glowing hydrogen already outside and within the visible globe of a sun.

On the whole it seems altogether more probable that the accession of splendour observed in the case of temporary stars is due to the downfall of enormous masses of matter upon the surface of these suns. It is, no doubt, well known to most of my readers that the downfall of meteoric matter upon the surface of our own sun has been considered a sufficient explanation of the sun's entire emission of light and heat. The theory that the sun's heat and light *are* thus excited has long since been abandoned; but not because the cause would be insufficient. It has been abundantly proved that a downfall òf meteors, not sufficient in quantity to add appreciably to the sun's size in many thousands of years, would generate more heat and light than he emits in that time. The meteoric theory has been abandoned simply because it has been shown that no such downfall is taking place.

The reason why meteoric impact would suffice to warm the sun to his present temperature if the meteoric showers were heavy, and to warm him far beyond his present temperature if for a few days very heavy meteoric showers fell upon him, is simply that his attraction upon matter approaching him from without is capable of generating a tremendous velocity. We know that when a cannon-ball strikes a metal target, with a velocity perhaps of some 400 yards per second, great heat is excited, and there is a momentary flash of light. If the velocity were doubled, the quantity of heat would be doubled also. Conceive, then, the tremendous heat which would be excited if a cannon-ball could be caused to strike a target with a velocity exceeding that just named some 1500 times! The ball and target would both be vaporised by the shock, if—which, however, could never happen—the target resisted the blow and brought the ball to rest. Now matter which reaches the sun from without, under the influence of his tremendous attraction, strikes his globe with a velocity 1500 times greater than that of a cannon ball striking a target at a distance of two or three hundred yards. The heat excited is, therefore, very intense; and if meteors were showering at all times and in dense flights upon the sun's surface, we should require no other explanation of the sun's heat.

But it appears that meteoric systems are neither so numerous nor so rich as to account for the sun's uniform emission of heat, though occasional meteoric showers upon the sun may be heavy enough to increase appreciably the amount of heat he emits. It would seem, from experiments which have been made by Professor Piazzi Smyth, of the Edinburgh Observatory, and later by the Astronomer Royal at Greenwich, that from time to time the sun's emission of heat really is greater than usual. It seems not at all improbable that the increase is due to the occasional fall of large masses of meteors in great numbers upon the sun.

Again, it seems that such falls occur periodically, or rather that at regular intervals great meteoric streams pour upon the sun's surface. For instance, the periodic increase and decrease in the number of sun-spots is accompanied (so far as we can judge by the observations made at Edinburgh and Greenwich) by an accession and diminution of the solar heat; and if the change is attributed to the passage of a meteoric stream athwart the sun, we should have to assign to such a stream a period of rather more than eleven years. This, from what we know about the association between meteors and comets, would correspond simply to the existence of a comet whose path intersects the sun's globe, and which is followed by a train of millions of large meteoric masses, many of which are consumed at each passage of the rich portion of the train athwart the globe of the sun. This comet must of necessity be inconspicuous, since it has hitherto escaped detection,

In fact, its head and nucleus must long since have been entirely destroyed. Only the meteoric train, far more widely scattered, remains, simply because at each passage past the sun, though many are captured, far greater numbers get safely past.

I am careful to remind the reader that though I have, for convenience, used the indicative mood in describing these matters, I am in reality presenting merely a theory. It may be that the solar spots and the accessions of heat are produced in some other way. But I must admit I find strong reasons for regarding as probable the general theory, that the alternations of solar activity (not the solar activity itself be it noted) are excited from without. And since we know, as a matter of fact, that meteors exist in enormous numbers within the solar system, and that they aggregate with rapidly increasing density in the sun's neighbourhood, we must believe that they fall upon the sun in enormous numbers. We also perceive that the supply cannot be uniform, but must vary greatly from time to time; while what we know about the periodicity of meteoric showers on our own earth suggests the belief, we may almost say the certainty, that there must be periodic downfalls of very heavy meteoric showers upon the sun's surface. We have, then, strong probability in favour of the belief that events may occur which, *if* they occurred, might be expected, with a high degree of probability, to produce effects resembling those actually observed,—viz., the production of a heat more intense than usual, accompanied by signs of great disturbance like the sun-spots. It does, therefore, seem at least not improbable that these accessions of heat and these signs of great disturbance really are brought about in the way supposed.

A further argument in favour of the meteoric origin of solar alternations of heat is to be found in the fact that, on one occasion at least, a solar phenomenon, corresponding precisely to what we should expect to see if great meteoric masses fell upon the sun, has been followed by precisely the same signs of terrestrial disturbance which accompany and follow the formation of great solar spots. I refer to the remarkable occurrence witnessed by Carrington and Hodgson (at different observatories) in September, 1859, when two intensely bright points of light were seen travelling beside each other at the rate of about 120 miles per second along a short arc of the sun's surface,—an arc only equal in length to some four-and-a-half times the diameter of our earth.

On that occasion the emission of solar heat may or may not have been increased in an appreciable degree for several minutes. My own belief is that it must have been; but we certainly have no means of proving that it was. What we do know certainly is, that on that day all the phenomena which usually accompany the existence of many and large sun-spots showed themselves with exaggerated intensity. The magnetic needle was greatly disturbed, auroras displayed their coloured streamers in both hemispheres, telegraphic communication was interrupted, and everything tended to show that a disturbance of the same general character as that which produces sun-spots, but much more active while it lasted, had affected the sun. It seems, then, altogether reasonable to infer that sun-spots are due to the same cause as the disturbance which then occurred. So that if we conclude, with most astronomers competent to form an opinion, that the disturbance witnessed by Carrington and Hodgson was due to the downfall of two very large meteoric masses upon the sun, it would follow that sun-spots are due to more wide-spread meteoric showers, not consisting of masses so large.

The reader will long since have guessed, no doubt, to what all this tends. If the periodical variations of the sun's surface are due to meteoric and cometic systems whose orbits intersect the sun's globe, their periods being short (that is, lasting but a few years), it may well be that more important meteoric and cometic systems intersecting the sun's globe exist, which have much longer periods. When next one of these makes its passage athwart the sun, far more important solar disturbances may take place than those which occur when the regularly recurring systems salute the sun. Two or three times in the history of science comets have approached very close to the surface of the sun, as in 1680, and again in 1843, but without actually impinging upon it. Very slight changes in the motions of those comets, owing to the disturbing influences of the planets, would cause their very nuclei to strike the sun, and their meteoric trains to pour afterwards in a full stream upon him for many days, or even for many months and years in succession.

Now I do not think our sun would necessarily suffer very much from any of these known comets. They may long since have parted with the greater quantity of their substance. But it is quite possible that even one of those well-known comets of the solar system might cause very serious outbursts of solar heat and light; and it is certainly not only possible but extremely probable that other comets, such as have visited the solar system on paths fortunately not bringing them near to the sun, would have worked much mischief had their paths been differently situated.

We know that Newton held this opinion. He considered the real danger from comets to reside, not in the possibility that one might strike our earth, but in the possibility that one, falling upon the sun, might excite that orb to a degree of heat so intense that all life on this earth would be destroyed. It is true that, in Newton's time, physical laws were not so well understood as at present, and a considerable portion of Newton's reasoning was consequently inexact. But nothing which is now known opposes itself to the belief which Newton adopted on this subject. On the contrary, whereas Newton only recognised the danger arising from the consumption of a comet as fuel for the sun, we now recognise a far more serious danger, from the force of meteoric impact, and the heat excited as the thermal equivalent of the destroyed velocities. Of this part of the danger Newton had no clear conception, the relations between mechanical energy and heat not having been established until quite recent times.

It appears to me, however, that the danger in the case of our own sun—or may we not say *our* danger?—arises only from the possibility that some one of the comets which visit us from the star-depths may make straight for the sun; and this danger is exceedingly small. Almost certainly a comet which, leaving the domain of another sun, falls under the attractive influence of our own, would approach him on a path passing many millions of miles from his surface. The chances against a more direct approach are so great that they may be regarded as, to all intents and purposes, overwhelming. A comet *might* visit us from the star-depth on a destructive course, just as a single black ball *might* be drawn at the first trial from a bag containing a million white balls and only that single black one. But the danger is exceedingly small.

We see, indeed, that other suns have suffered in this way, assuming cometic downfall to be the true cause of stellar outbursts. There are so many millions of suns, however, in the region of space to which telescopic survey extends that the occurrence of ten or twelve such outbursts in the course of four or five centuries need not be regarded as implying any serious danger. Moreover, all the suns which have thus suffered lie within a particular region of the heavens,—viz., in the Milky Way, and in that half of the Milky Way which is most irregular, one may almost say *ragged*, in structure. (With one exception—the star in the Northern Crown, which, nevertheless, lies on a faint outlying streamer of the Milky Way not discernible to ordinary vision.) If then our sun belongs to this region of space, the danger for him and for us is somewhat greater than my previous argument would indicate. For, in that case, we must compare the number of outbursts, not with the total number of stars within telescopic range, but with the number of those stars which lie within this particular region of space. On the other hand, if our sun does not lie within that region of space, the danger for him and for us is very much less; for instead of a certain small number of accidents among his fellow suns, there have been no such accidents, only accidents affecting other suns which must be differently classed.

The case may be compared to the estimation of the dangers, let us say, of travelling by ocean steamships on a particular route. If we take the total number of accidents, for instance, to steamships travelling between England and the United States, we should estimate the risk of the journey as very small, the number of passengers who have lost their lives being very small compared with the number who have made the journey. But even this small risk is diminished if we estimate the danger for a passenger by Cunard steamships, simply because no passenger has yet lost his life through accident to one of these Cunard vessels.

So in the case of our sun, the danger of an outburst such as has affected the stars in the Northern Crown and Cygnus is small enough when we estimate it by comparing the number of such accidents with the total number of stars, but vanishes almost into nothingness when we note that no insulated star like our sun seems hitherto to have undergone one of these tremendous catastrophes.

But as regards the fate of worlds circling round suns which have suffered in this way, we can form but one opinion. Beyond all doubt, if such worlds existed and were inhabited when their central orb blazed forth with many hundred times its former lustre, all life must have perished from their surface. We may believe, as many do, that no conditions are too unlike those we are familiar with on earth to render life impossible; that the creatures subsisting in a world exposed to the most fiery heat or to the most intense cold are adapted as perfectly to the conditions under which they subsist as we are to the circumstances of terrestrial life. But even adopting this view, though it seems to accord ill with what we know of our own earth,—where life ceases towards the polar and over large tracts of the equatorial regions,—we could not believe that creatures thus adapted to the conditions prevailing around them could endure an entire change of those conditions. With the accessions of heat in the stars in Cygnus and the Crown, such change must inevitably have taken place. Therefore, as I think, we must regard the catastrophes affecting those remote suns as assuredly involving " The End of many Worlds."

Note.—What is stated in the latter portion of this chapter applies now only to the star in the Northern Crown ; for the star in Cygnus has not faded into a small star, but into a small nebula ! For the further history of this star, the reader is referred to my forthcoming treatise entitled, "Pleasant Ways in Science."

VI.

THE AURORA BOREALIS.

AMONG the objects in view, when the recent Polar expedition was fitted out, was the hope that during the winter of 1875-76 the scientific observers who accompanied the expedition might be able to study the Aurora Borealis under unusually favourable conditions. This hope was, as most of my readers doubtless know, disappointed. Few auroras were seen, and those seen were not remarkable either for brilliancy or for beauty of colour. Yet in the very disappointment of the hope which had been entertained on this subject there was very significant evidence respecting the aurora, as will presently be shown. The quiescence, at that time, of the forces which produce the auroral streamers had its meaning, and a very strange one.

The aurora is one of those phenomena of nature which are characterized by exceeding beauty, and sometimes by an imposing grandeur, but are unaccompanied by any danger, and indeed, so far as can be determined, by any influence whatever upon the conditions which affect our well-being. Comparing the aurora with a phenomenon akin to it in origin—lightning—we find in this respect the most marked contrast. Both phenomena are caused by electrical discharges; both are exceedingly beautiful. It is doubtful which is the more imposing so far as visible effects are concerned. When the auroral crown is fully formed, and the vault of heaven is covered with the auroral banners, waving hither and thither silently, now fading from view, anon glowing with more intense splendour, the mind is not less impressed with a sense of the wondrous powers which surround us than when, as the forked lightnings leap from the thundercloud, the whole heavens glow with violet light, and then sink suddenly into darkness. The solemn stillness of the auroral display is as impressive in its kind as the crashing peal of the thunderbolt. But there is a striking contrast between the feelings with which we regard the safe splendours of the aurora and the terrible glory of the lightning flash. One display we contemplate with the calmness engendered by absolute security; the other —no matter how little the fear of death may affect the reason—cannot be regarded without exciting the consciousness of danger. We witness in safety, so far as itself is concerned, the flash whose light illuminates the cloud masses above and around us, but for aught we know it may be the last we shall ever see, since no man killed by lightning ever saw the flash which brought his death.

I do not purpose to consider here at any length those facts respecting the aurora which properly find their place in text-books of science, but those only which are less commonly dealt with, and seem at once most suggestive and most perplexing.

The reader is no doubt aware that auroras, or polar streamers, as they are sometimes called, are appearances seen not around the true poles of the earth, but around the magnetic poles, which lie very far away from those geographical poles which our arctic and antarctic seamen have in vain attempted to reach. We in England, though much nearer to the north pole than the inhabitants of Canada, see far fewer auroras than they do, and those we see are far less splendid, simply because we are farther away from the northern magnetic pole. This will be seen from the accompanying pair of maps (from my "Elementary Physical Geography"), showing where the northern and southern magnetic poles lie. Again, you will see from the northern map, that from England the northern magnetic pole lies towards the west of due north. That is why when we see a fully developed auroral arch in this country its crown lies towards the west of north (almost midway between north and north-west). I may have occasion at another time to consider the curious changes which affect the actual position of

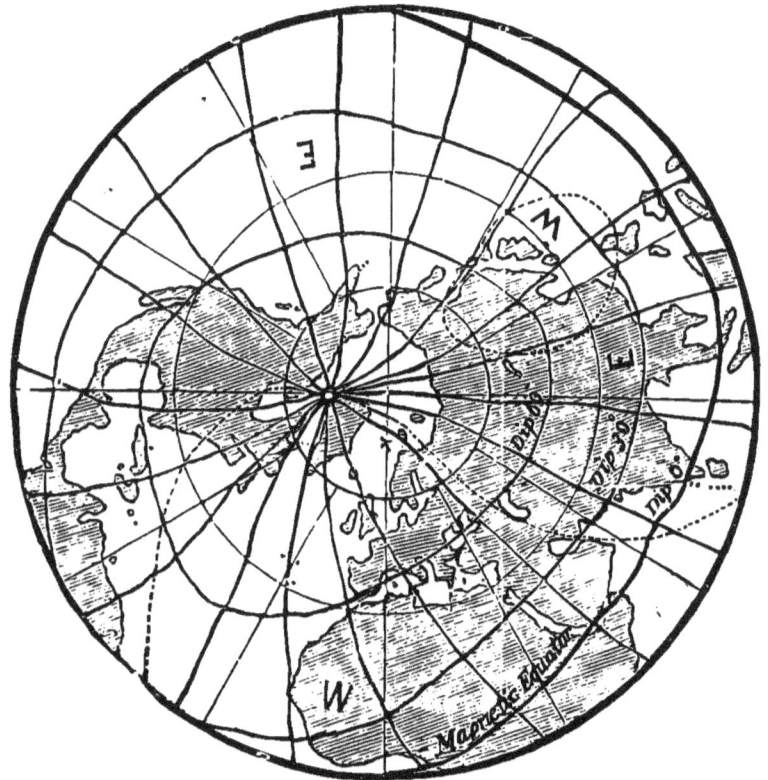

Fig. 9.—The Northern Magnetic Meridians and Lines of Equal Dip.

the magnetic poles and lines; in this place I merely note that what is now said respecting them only refers to the present time.

The formation of auroral streamers around the magnetic poles of the earth shows that these lights are due to electrical discharges, just as the general magnetic phenomena of the earth indicate the existence of electrical currents. The earth, in fact, with its envelope of air, moist and dense near the surface, rare and dry above

Fig. 10.—The Southern Magnetic Meridians and Lines of Equal Dip.

may be regarded as an enormous magnetic instrument, a core surrounded by conducting matter, in which electrical currents pass whenever the condition of the earth's magnetism changes. The discharges of electricity, though only visible at night, take place in reality in the daytime also. According to their extent and position, varying with the varying conditions under which they take place, their aspect changes. Moreover, from different parts of the earth the appearance of the aurora is different. From low latitudes (I speak now of magnetic latitudes as indicated by the closed curves around the magnetic poles in the maps), the auroral arch is seen towards the north in our hemisphere, towards the south in the other hemisphere. From points nearer the magnetic pole it is seen overhead, and when that pole is approached still nearer, the crown of the arch is seen on the side remote from the pole,—that is, towards the south in our hemisphere, towards the north in the southern hemisphere.

Remembering that the aurora is due to electrical discharges in the upper regions of the air, it is interesting to learn what are the appearances presented by the aurora at places where the auroral arch is high above the horizon,—these being, in fact, places nearly *under* the auroral arch. M. Ch. Martins, who observed a great number of auroras at Spitzbergen in 1839, thus writes (as translated by Mr. Glaisher) respecting them: "At times they are simple diffused gleams or luminous patches; at others, quivering rays of pure white which run across the sky, starting from the horizon as if an invisible pencil were being drawn over the celestial vault; at times it stops in its course, the incomplete rays do not reach the zenith, but the aurora continues at some other point; a bouquet of rays darts forth, spreads out into a fan, then becomes pale, and dies out. At other times long golden draperies float above the head of the spectator, and take a thousand folds and undulations as if agitated by the wind. They appear to be but at a slight elevation in the atmosphere, and it seems strange that the rustling of the folds as they double back on each other is not audible. Generally, a luminous bow is seen in the north; a black segment separates it from the horizon, the dark colour forming a contrast with the pure white or bright red of the bow, which darts forth rays, extends, becomes divided, and soon presents the appearance of a luminous fan, which fills the northern sky, and mounts nearly to the zenith, where the rays, uniting, form a crown, which in its turn darts forth luminous jets in all directions. The sky then looks like a cupola of fire; the blue, the green, the yellow, the red, and the white vibrate in the palpitating rays of the aurora. But this brilliant spectacle lasts only a few minutes; the crown first ceases to emit luminous jets, and then gradually dies out; a diffused light fills the sky; here and there a few luminous patches, resembling light clouds, open and close with incredible rapidity, like a heart that is beating fast. They soon get pale in their turn, everything fades away and becomes confused, the aurora seems to be in its death-throes; the stars, which its light had obscured, shine with a renewed brightness; and the long polar night, sombre and profound, again assumes its sway over the icy solitudes of earth and ocean."

The association between auroral phenomena and those of terrestial magnetism has long been placed beyond a doubt. Wargentin in 1750 first established the fact, which had been previously noted, however, by Halley and Celsius. But the extension of the relation to phenomena occurring outside the earth—very far away from the earth—belongs to recent times.

The first point to be noticed, as showing that the aurora depends partly on extra-terrestrial circumstances, is the fact that the frequency of its appearance varies greatly from time to time. It is said that the aurora was hardly ever seen in England during the seventeenth century, though the northern magnetic pole was then much nearer to England than it is at present. Halley states that before the great aurora of 1716 none had been seen (or at least recorded) in England for more than eighty years, and no remarkable aurora since 1574. In the records of the Paris Academy of Sciences no aurora is mentioned between 1666 and 1716. At Berlin one was recorded in 1707 as a very unusual phenomenon; and the one seen at Bologna in 1723 was described as the first which had ever been seen there. Celsius, who described in 1733 no less than three hundred and sixteen observations of the aurora in Sweden between 1706 and 1732, states that the oldest inhabitants of Upsala considered the phenomenon as a great rarity before 1716. Anderson, of Hamburg, states that in Iceland the frequent occurrence of auroras between 1716 and 1732 was regarded with great astonishment. In the sixteenth century, however, they had been frequent.

Here, then, we seem to find the evidence of some cause external to the earth, as producing auroras, or at least as tending to make their occurrence more or less frequent. The earth has remained to all appearance unchanged in general respects during the last three centuries, yet in the sixteenth her magnetic poles have been frequently surrounded by auroral streamers; during the seventeenth these streamers have been seldom seen; during the last two-thirds of the seventeenth century auroras have again been frequent; and during the present century they have occurred sometimes frequently during several years in succession, at others very seldom.

Let us inquire a little more closely into the circumstances attending auroral displays, in order to ascertain what external cause it is which thus influences their occurrence,

Connected as auroras are with the phenomena of

terrestrial magnetism, we may expect to find some help in our inquiry from the study of these phenomena.

Now it appears certain that magnetic phenomena are partly influenced by changes in the sun's condition. We may well believe that they are in the main due to the sun's ordinary action, but the peculiarities which affect them seem to depend on *changes* in the sun's action. It is found that the daily oscillation of the magnetic needle corresponds with the diurnal change in the position of the sun owing to the earth's rotation. An annual change affecting that oscillation depends on the varying distance of the sun as the year proceeds. The daily change is not only greater than the annual, but is characterized by irregularities, when the face of the sun shows the greatest number of spots. It was found by General Sabine, says Mr. Balfour Stewart, "that the aggregate value of magnetic disturbances at Toronto attained a maximum in 1848, nor was he slow to remark that this was also Schwabe's period of maximum sun-spots. It was afterwards found, by observations made at Kew, that 1859 (another of Schwabe's years) was also a year of maximum magnetic disturbance. . . . There is also some reason to believe that on one occasion our luminary was caught in the very act. On the first of September, 1869, two astronomers, Carrington and Hodgson, were independently observing the sun's disc, which exhibited at that time a very large spot, when, about a quarter past eleven, they noticed a very bright star of light suddenly break out over the spot and move with great velocity across the sun's surface. On Mr. Carrington sending afterwards to Kew Observatory, at which place the position of the magnet is recorded continuously by photography, it was found that a magnetic disturbance had broken out at the very moment when this singular appearance had been observed." The dip of the magnetic needle, its deflection from the north, the inferiority of its directive force, were all three simultaneously and abruptly altered, and continued so for many hours.

Nor are we left in any doubt as to the connection between such well-marked disturbances of the magnetic needle. While the needle was thus violently displaced, vivid auroras occurred over the greater part of both the northern and southern (magnetic) hemispheres. They were seen in latitudes where usually auroras are as infrequent as rain in Peru,—at Rome, in the West Indies, even within eighteen degrees of the equator.

The disturbance of the earth's electrical condition was well shown in other ways. Mr. C. V. Walker, the telegraphist, found that strong electrical currents affected the various telegraphic lines throughout England. These currents changed in direction every two or three minutes. In many places it was impossible to send telegraphic messages. In America some of the signalmen received severe electric shocks. "At a station in Norway," says Sir J. Herschel, "the telegraphic apparatus was set fire to; and at Boston, in North America, a flame of fire followed the pen of Bain's electric telegraph (which writes down the message upon chemically prepared paper)."

Many of my readers will doubtless remember the auroras of May 13, 1869, and October 24, 1870, both of which occurred when the sun's surface was marked by many spots, and both of which were accompanied by remarkable disturbance of the earth's magnetism.

It may, then, fairly be assumed that the occurrence of auroras depends in some way, directly or indirectly, on the condition of the sun. But what the real nature of that connection may be is not to be easily determined. It is clear that the eleven-year-period of sun-spots is not the only, or even the chief period affecting auroras, for we have seen that sometimes for a full century, or even more, very few auroras are seen. It is not by any means certain that the connection between the sun's condition and the occurrence of auroras is of the nature of cause and effect; quite probably sun-spots and auroras depend on some common cause as yet undetected,—and possibly never to be detected by man.

Regarding the auroral streamers as terrestrial lights only, but in some sense like the light reflected by planets in having their real source in the sun, we can no longer speak, as Humboldt was wont to do, of our planet possessing a power of emitting light of its own. Yet his manner of dealing with auroral light still possesses interest for us, especially in relation to the question whether these polar lights are emitted by other planets and may possibly be discerned from our earth. "It results from the phenomena of the aurora," said Humboldt, "that the earth is endowed with the property of emitting a light distinct from that of the sun. The intensity of this light is rather greater than that of the moon in its first quarter. It is at times, as on January 7, 1831, strong enough to admit of one's reading printed characters without difficulty. This light of the earth, the emission of which towards the poles is almost continuous" (this, however, is not strictly the case), "reminds us of the light of Venus, the part of which not lighted by the sun often glimmers with a dim phosphorescent light. Other planets may also possess a light evolved out of their own substance."

I would venture, however, to express strong doubts as to the possibility of discerning, either on Venus or on any other planet, the auroral gleams which may very probably illuminate at times their nocturnal skies. It must be remembered that the aurora, when at its brightest and covering a large part of the sky, only gives about as much light as the moon in her first quarter,—that is, as one half of a disc so small that 180,000 such discs would not equal the entire sky. The luminosity of the aurora is then in reality very small; probably far less than that of the earth's surface when illuminated by the full moon. A distant hill on which the rays of the full moon are falling seems strongly illuminated, and yet its light is really so faint that we could scarcely discern it at all save for the favouring effect of contrast. We know this, because we often see portions of the moon's surface which are illuminated by earthshine (when we see what is called the old moon in the new moon's arms), and these portions are quite faint by comparison with the rest of the moon; yet earthshine exceeds moonshine at least twelve times, and probably more nearly twenty times in splendour.

The glimmering phosphorescent light, supposed to have been seen on parts of Venus not lighted by the moon, is a phenomenon about which experienced telescopists are somewhat doubtful, though Webb speaks of the appearance as remarkably well attested, quoting, amongst others, the following cases. In 1715, Derham, in his "Astro-Theology," says that " the sphericity or rotundity is manifest in our moon, yea, and in Venus, too, in whose greatest falcations " (*i.e.*, when they appear as crescents) "the dark parts of their globes may be perceived, exhibiting themselves under the appearance of a dull and rusty colour." In 1806, the phenomenon displayed itself beautifully to Harding three times and to Schröter once within five weeks. "Guthrie and others noticed it a few years ago, with small reflectors, in Scotland; Purchas, at Ross, in England; De Vico and Palomba, many times in Italy." Winnecke records a similar observation, though very faint, 1871, September 25, a little before noon. Van Hahn also says he saw it repeatedly, by day as well as by night, and with several instruments; he was, however, an inferior observer. The dark side is sometimes described as grey, sometimes as reddish. The phenomenon has, on the other hand, been looked for specially, on several occasions, by practised observers, using very fine instruments, who have failed to recognise any trace of it.

One of the most remarkable observations ever made on Venus must here be mentioned. Mädler states that on one occasion, when he was observing the planet, he saw a number of brushes of light diverging from the circular side (*i.e.*, the outside of the planet's crescent), lasting as long as the planet could be seen that evening, and remaining unchanged when he changed the position of the telescopic eye-piece, or used a different one. "He attempts no explanation," says Webb, "but thinks it could not have been an optical illusion. This is certainly *possible*, but it is an instructive instance of the oversights which may be incidental even to great philosophers, that it never seems to have occurred to him to try another telescope!" It cannot be doubted that the evidence would have been greatly strengthened had he changed telescope as well as eye-piece; though it is not readily to be explained how a known telescope, frequently used as well before as after this strange appearance was seen, could for one evening only have played so strange a trick as Mädler's must have done, if what he saw was merely an instrumental illusion.

However, whether we have telescopic evidence or not respecting auroral lights surrounding the polar regions of other planets, we can have very little doubt that some among the planets, if not all of them, resemble our earth in this as in so many other respects. The aurora is a cosmical phenomenon, not one peculiar to our own earth. It is not, indeed, altogether certain that our sun himself may not be girt round by mighty auroral streamers, and that the light of these may not constitute a noteworthy portion of the corona of glory seen around him during the time of total eclipse.

This view, indeed, although it has not been definitely entertained as I have here expressed it, has been suggested by reasoning which led others to suppose that the coloured prominences around the sun may be auroras. Perceiving the nature of the connection between terrestrial magnetism and auroras, Balfour Stewart reasoned that we may extend our inquiries and ask, "If the sun's action is able to create a terrestrial aurora, why may he not also create an aurora in his own atmosphere?" It occurred independently to General Sabine, Prof. Challis, and himself, that the red flames visible during a total solar eclipse "may be solar auroræ." We now know that the solar flames are not auroræ, nor, properly speaking, flames at all, but great masses of glowing vapour. It is not, however, by any means so clear that the solar corona is not auroral in its nature. The following reasoning, applied by Balfour Stewart to the sun's prominences, applies with much greater force to the corona. After mentioning the height (from 70,000 to 80,000) which some prominences attain, he proceeds, "Considering the gravity of the sun, we are naturally unwilling to suppose that there can be any considerable amount of atmosphere at such a distance from his surface; and we are therefore induced to seek for an explanation of these red flames amongst those phenomena which require the smallest possible amount of atmosphere for their manifestation. Now the experiments of Mr. Gassiot and the observed height of the terrestrial aurora alike convince us that this meteor will answer our requirements best. And besides this, the curved appearance of these red flames, and their high actinic power, in virtue of which one of them, not visible to the eye, was photographed by Mr. De la Rue, are bonds of union between these and terrestrial auroræ."

All this and much more may be said of the solar corona. Its streamers extend not 70,000 or 80,000 miles, but 700,000 or 800,000 miles from the surface of the sun, where the pressure must be far smaller than near the summits of even the loftiest prominences. They are curved and striated, like those of the aurora, whereas the shapes of the prominences bear only a distant resemblance to auroral streamers. They possess a high actinic (*i.e.*, photographic) power, as is shewn by the readiness with which, during the total eclipse of December, 1871, they were photographed, no less than six well-defined negatives being taken both by Col. Tennant, at Ootacamund, and by Mr. Davis, at Baikal, during the brief continuance (only a few minutes) of total obscuration. In every respect the solar corona accords far better than do the solar coloured prominences with the appearance we should expect to recognise in solar auroras.

In particular, it has always seemed to me that the curved, especially the doubly curved, streamers of the corona can only be well explained by regarding the corona as in the main an auroral phenomenon. If mighty currents prevailed in the higher regions of a rare atmosphere, extending hundreds of thousands of miles from the sun's surface, appearances such as these curved streamers would undoubtedly be explained. But no one who considers the effect of the sun's tremendous attractive power on such an atmosphere can fail to perceive that, according to the known laws connecting gaseous pressure and density, the density of that atmosphere would be enormously great, even at a very great distance from the sun's surface, if the curved streamers really were caused by atmospheric currents. We know, on the contrary, from the behaviour of comets which have passed very near to the sun, that the atmosphere above his visible surface must be very rare indeed.

It must not be understood, however, that I regard the corona as simply a great solar aurora. It is certain that the whole region filled by the corona is occupied by immense numbers of scattered meteors, and extremely probable that large quantities of cometic matter exist within the same region. Vaporous masses may also be there, circling independently around the sun. But that this region is illuminated constantly by auroral light, varying greatly in intensity and position, seems very strongly indicated by all that we know about the corona, as seen during different total eclipses of the sun.

If we so viewed the solar corona, and found our earth, therefore, in this respect resembling the great central orb of the solar system, we could not but regard as extremely probable the theory that other planets also resemble the central body in this respect. We might then picture to ourselves every orb in the solar system carrying onward its faintly luminous crowns of boreal and austral light, not shining with constant lustre, or in the same constant position, but at one time leaping in coloured steamers to a great distance from the body they adorned, and anon sinking down and growing fainter and fainter, or occasionally disappearing altogether. Then, when some great disturbance affected the central sun, and caused *his* auroral banners to shine out more brilliantly and to attain a greater extension, suddenly the auroral streamers of all the planets would leap out into new light and life, playing around the northern and southern magnetic poles of those orbs, even as electric brushes play around the positive and negative electrodes of a Geissler's tube. "Suddenly" at least so far as each planet is concerned, but not suddenly throughout the whole system. For the magnetic influences, like the light and heat of the sun, require time for their transmission. Yet, so rapidly do they travel that, in a few hours, the auroral illumination would extend from the central sun to the outermost limits of his system.

It remains that I should make a few remarks on the evidence which that wonderful instrument of research, the spectroscope, has afforded respecting the light of the aurora.

Angström was the first to observe the spectrum of the aurora borealis. He found that the greater part of the auroral light, as observed in 1867, was of one colour, yellow, but three faint bands of green and greenish blue colour were also seen. The aurora of April 15, 1869, was seen under very favourable conditions in America. Prof. Winlock, observing it at New York, found its spectrum to consist of five bright lines, of which the brightest was the yellow line just mentioned. One of the others seems to agree very nearly, if not exactly, in position with a green line, which is the most conspicuous feature of the spectrum of the solar corona. During the aurora of October 6, 1869, Flögel noticed the strong yellow line and a faint green band. Schmidt, on April 5, 1870, made a similar observation. He saw the strong yellow line, and from it there extended towards the violet end of the spectrum a faint greenish band, which, however, at times showed three defined lines, fainter, than the yellow line.

It was not till the magnificent aurora of October 24, 25, 1870, that any red lines were seen in the spectrum of an aurora. On that occasion the background of auroral light was ruddy, and on the ruddy background there were seen three deep red streamers very well defined. The ruddy streamers, on the night of October 25, converged towards the auroral crown, which was on that occasion singularly well seen. Förster of Berlin failed to see any red line or band despite the marked ruddiness of the auroral light. But Capron at Guildford saw a faint line in the red part of the spectrum; and Elger at Bedford observed a red band in the light of the red streamers, the band disappearing, however, when the spectroscope was directed on the white rays of the aurora.

As yet the auroral spectrum has not been interpreted. It is not a spectrum which can be (at present) artificially produced. We understand the spectrum of the sun and stars, because spectra of the same order can be produced in our laboratories. The spectra of the planets, so far as they differ from the spectrum of reflected sunlight in showing signs of the absorptive action of the planetary atmosphere, have been similarly interpreted. So also the spectra of the coloured solar prominences are understood, while those of nebulæ and comets, though not as yet thoroughly explained, have been partly interpreted, because of their partial agreement with the known spectra of earthly elements. But as yet neither the spectrum of the aurora nor that of the solar corona has been explained. The reason probably is, that the conditions under which the light of the aurora as of the corona is formed are not such as have been or perhaps can be attained or even approached in laboratory experiments.

VII.

THE LUNAR HALO.

THERE are some phenomena of nature which suggest false ideas. For instance, when we look at the broad expanse of ocean on a moonlit night, and see a path of glory on its surface, directed towards the moon's place, we seem to be assured by the sense of sight that that broad track is illuminated while the waters all around are dark. A little consideration, however, assures us that the impression is a false one, that in this case seeing is not believing. The moon's rays really illumine the whole surface which lies before us, and we fail to receive light from other parts than the track below the moon, *not* because they receive no light, but because the light which they receive is not reflected towards us. An observer, stationed a mile or two towards the right or towards the left of our station, sees a different track of light, while the part which seems bright to us seems dark to him.

The rainbow is another phenomenon of this deceptive kind. We seem to see an arch of many colours suspended in the air,—and when we learn that it is due to the presence of drops of water in the air, we are apt to infer that where we see the red arch there are drops lit up with red light, where the yellow, green, or violet arch, that the drops are aglow with yellow, green, or violet light. But in reality this is not so; the same drops which seem green to us will seem red to another observer, violet to another, and to yet other observers will show none of the prismatic colours, but only the dull grey colour of the cloud on which the rainbow is seen. We have here a pretty emblem of the varied aspects which events of the same real nature present to different persons, or according to the different circumstances under which the same person may see them. One shall see events in rosy tints, or with the freshness of spring hues, or with the melancholy symbolled by the

> deeper indigo (as when
> The heavy-skirted evening droops with frost)—

while to others the same events shall show only the ordinary tints of common-place life.

The lunar halo is one of the phenomena thus deceptive to the view. We see all around the moon a circle or arc of light, nearly white, though sometimes faint tints of colour can be perceived in it, while the space within the circle seems manifestly darker than the space outside. The appearance of the halo as seen under favourable conditions is shown in fig. 11, on the next page. In this country the dark space round the moon is not generally so well seen as in countries where the air is clearer. But this is in reality the characteristic feature of the halo, as its name shows. For the name is derived from a Greek word signifying threshing-floor (the old threshing-floors being round), and thus naturally describes a round space relatively clear, surrounded on all sides by a ring of aggregated matter.

We seem in looking at the lunar halo, then, to see the moon at the centre of a dark space, surrounded by a ring of bright particles, outside which again are particles not quite so brightly illuminated as those forming the ring, but more brightly than those within the ring.

But in reality this impression, which, so far as the sense of sight is concerned, seems *forced* upon the mind, is entirely erroneous. There is no real distinction between the space which looks dark all round the moon, the space beyond which does not look dark, and the ring between the two spaces which looks bright. These are all equally illuminated by the moon, in the same sense, at least, that we say the surface of a moonlit sea is all equally illuminated, neglecting slight differences which do not concern the point we are specially dealing with. Precisely as the path of light on the ocean is not a real path of illumination, bounded on either side by

Fig. 11.—Lunar Halo

dark spaces, so the ring of light round the moon is not a real ring of light, bounded on one side by a less bright region, and within by a dark space.

Although my object in these essays is not specially to

deal with scientific matters, but rather with the thoughts (much more important in my belief) which they suggest—so that, in dealing with my present subject, I wish rather to call attention to the manifold ways in which our senses may deceive us unless their evidence is carefully cross-examined—yet it may be worth while to notice how the particular illusion here considered has deceived even the scientific elect.

It had been noticed by Tyndall, in certain experiments, that a very sensitive measurer of heat, when placed under the moon's rays, gathered together by a powerful condenser, seemed to indicate cooling rather than heating, as we should expect. On this a French student of science pointed to the darkening under the moon where the lunar halo is seen as evidence that our satellite possesses a certain power of clearing away vaporous matter from the air. "*On peut dire*," he said, speaking of the dark space within the halo, "*que la lune ouvre alors une porte par laquelle s'échappe le calorique que l'action solaire a emmagasiné dans les couches inférieures.*" "One may say," that is, "that the moon then opens a door through which the heat escapes, which the sun's action has stored up in the lower layers" (of the air). It will be manifest, if we remember that a lunar halo can often be seen at the same time from stations hundreds of miles apart, that there can be no such opening of clear air,

For the cloud layer in which the halo is formed is but a few miles above the observer; and therefore, if one observer saw a circular opening in this layer, with the moon at its centre, another, a hundred miles from him, would see the space in a very different direction. The moon would not only not be at the centre of the space for this second observer, but would not be visible through the space at all. Moreover, the space could not possibly seem round to both observers; if it seemed round to one, it would look like a very flat oval of darkness (almost a mere line) to the other.

The real explanation of the lunar halo is very different. When you see such a halo, you may be certain that there is, high up in the air, a layer of light feathery cloud—the cirrus cloud, as it is called—composed of tiny crystals of ice. These crystals, as we know from those which in winter sometimes fall (not as snow, but as little ice-stars), have all a definite shape. They are in fact little prisms of ice, with angles like those of an equilateral triangle. These little prisms deflect the light which falls upon them, just as one of the drops of a chandelier deflects any light which falls upon it. If you hold a prism-drop of a chandelier between the eye and a light, you will see that the prism looks dark; it is really lit up, but it sends the light away in such a direction that the eye receives none. Now move it gradually away from the line of sight to the light, and at a certain distance it appears full of light; or, to speak more correctly, it sends the light it receives directly towards your eye. Beyond that position it again looks dark, but not so dark as when it was nearly between the eye and the light.

The little crystals of ice perform the same part with respect to the moon, when we see a lunar halo. Those between us and the moon, or within a certain distance from the line of sight to the moon, are, in reality, lit up by the moon's rays; but they send off those rays in such directions that we do not receive the light. Thus, all the space lying towards the moon, and for a certain distance all round, looks dark. But, at a certain distance, these little crystals send us light. If we could see them separately, they would seem to be full of light. That is the distance where ice-crystals of their known shape act most favourably in deflecting light,—that is, send off most for all the varying positions (not places) they can be in. At greater distances, a small proportion send us light. Thus, at that distance we have a ring of light, and outside the ring we have a gradual falling off in the quantity of light.

But the reader will be apt, perhaps, to say, How can all this be proved? No one has ever been among the ice-crystals of the feathery clouds when they are performing this work. When Coxwell and Glaisher made their highest ascent, the feather-clouds seemed almost as high above them as ever. Nor, if any one could reach those clouds, could he see the ice-crystals at their work. Yet there are few points about which science is more certainly assured than about this explanation of the halo. For we know the shape constantly assumed by ice-crystals; we know according to what precise law ice bends rays of light falling upon it; hence we can calculate quite certainly where, if ice-crystals make the halo, its rings should be seen. And the halo has the precise position thus calculated from the known laws of optics, and the known facts about ice and ice-crystals. The diameter of the halo should be, and is, about eighty times the apparent diameter of the moon, or somewhat less than half the arc which separates the point overhead from the horizon.

There is, however, yet stronger evidence. Haloes form around the sun as well as round the moon,—in fact, more frequently. Solar haloes have so much more light in them that we can recognise varieties of tint. Now, it follows from the laws of optics that, for the red part of the sun's light, the halo ring should have a smaller diameter than the halo ring for the violet part, intermediate colours having their corresponding intermediate halo rings. Thus, the halo ring, as a whole, should be rainbow-tinted, red on the inside, then orange, yellow, green, blue, indigo, and violet; and these colours are shown (under favourable conditions) in this order.

The student looking out for haloes, solar or lunar, must be careful not to confound them with solar and lunar coronas, that is, not the corona of astronomy, but rings of light around the sun and moon, much smaller than the true halo rings. What I have said above about the size of the true halo will suffice to prevent such a mistake. Coronas are not nearly so *easily*, though they have been quite as thoroughly, explained by science, as haloes.

It is singular to observe how utterly unlike the interpretation of the halo by science is from the natural interpretation. The observer would say, There surely is a dark space all round the moon, and round that a ring of light,—I see these things, and seeing is believing. Science says there is no dark space, and there is no ring of light; while the eye of science perceives something where the lunar halo shines which ordinary vision cannot recognise. Up yonder, many miles above the earth, science sees millions of crystals of ice, carried hither and thither—so light are they—by every movement of the air. Science sees these ice crystals deflecting the rays of moonlight, sifting the red rays from the orange, and these from the yellow, yellow from green, green from blue, blue from indigo, and indigo from violet. Science, in fine, perceives processes taking place in those higher regions of air compared with which the most delicate analyses of the laboratory are utterly coarse and imperfect.

There is a purer and nobler poetry in the lunar halo as thus understood than in its mere visible phenomena, attractive and beautiful though these are. Idle indeed is the fear that the interpretation of this special mystery of nature will leave the number of nature's mysteries diminished by one. On the contrary, for the one mystery explained many deeper mysteries are suggested. The phenomena discernible by the sense of sight are explained, but only by bringing into the range of a purer and more piercing vision phenomena infinitely more wonderful. If one could see through some amazing extension of visual power, or if even the imagination could adequately picture, the rush of light waves of all orders of length upon the line of crystal breakers, their deflection in all directions, their separation into their various orders of wave-length; if one could perceive the actual illumination of the ice-crystals, even where they seem dark to us, and the continual fluctuations of the troubled sea of ether between the crystal breakers and the earth below,—the scene would infinitely transcend in interest and mystery, the picture would be infinitely more suggestive of solemn thoughts, than the scene—beautiful though it doubtless is—presented by the halo-girt moon to ordinary vision. Truly they know little of the real meaning of science who regard it as depriving natural phenomena of their effect on the imagination, as robbing Nature of her poetic influence.

VIII.

MOONLIGHT.

THE light of the moon and the changes of the moon were probably the first phenomena which led men to study the motions of the heavenly bodies. In our times, when most men live where artificial illumination is used at night, we can scarcely appreciate the full value of moonlight to men who cannot obtain artificial light. Especially must moonlight have been valuable to the class of men among whom, according to all traditions, the first astronomers appeared. The tiller of the soil might fare tolerably well without nocturnal light, though even he,—as indeed the familiar designation of the harvest-moon shows us,—finds special value, sometimes, in moonlight. But to the shepherd moonlight and its changes must have been of extreme importance as he watched his herds and flocks by night. We can understand how carefully he would note the change from the new moon to the time when throughout the whole night, or at least of the darkest hours, the full moon illuminated the hills and valleys over which his watch extended, and thence to the time when the sickle of the fast waning moon shone but for a short time before the rising of the sun. To him, naturally, the lunar month, and its subdivision, the week, would be the chief measure of time. He would observe—or rather he could not help observing—the passage of the moon around the zodiacal band, some twenty moon-breadths wide, which is the lunar roadway among the stars. These would be the first purely astronomical observations made by man; so that we learn without surprise that before the present division of the zodiac was adopted the old Chaldean astronomers (as well as the Indian, Persian, Egyptian, and Chinese astronomers, who still follow the practice) divided the zodiac into 28 lunar mansions, each mansion corresponding nearly to one day's motion of the moon among the stars.

It is easy to understand how the first rough observations of moonlight and its changes taught men the true nature of the moon, as an opaque globe circling round the earth, and borrowing her light from the sun. They perceived, first, that the moon was only full when she was opposite the sun, shining at her highest in the south at midnight when the sun was at his lowest beneath the northern horizon. Before the time of full moon, they saw that more or less of the moon's disc was illuminated as he was nearer or farther from the position opposite the sun, the illuminated side being towards the west—that is, towards the sun; while after full moon the same law was perceived in the amount of light, the illuminated side being still towards the sun, that is, towards the east. They could not fail to observe the horned moon sometimes in the daytime, with her horns turned directly from the sun, and showing as plainly, by her aspect, whence her light was derived, as does any terrestrial ball lit up either by a lamp or by the sun.

The explanation they gave was the explanation still given by astronomers. Let us briefly consider it. In doing so I propose to modify the ordinary text-book illustration which has always seemed to me ingeniously calculated (with its double set of diversely illuminated moons around the earth) to make a simple subject obscure.

In fig. 12, let E represent the earth one half in darkness, the other half illuminated by the rays of the sun S, which should be supposed placed at a much greater distance to the left,—in fact, about five yards away from E. To preserve the right proportions, also, the sun ought to be much smaller and the earth a mere point.

I mention this to prevent the reader from adopting erroneous ideas as to the size of these bodies. In reality it is quite impossible to show in such figures the true proportions of the heavenly bodies and of their distances. Next let M_1, M_2, M_3, etc., represent the moon in different positions along her circuit around the earth at E.

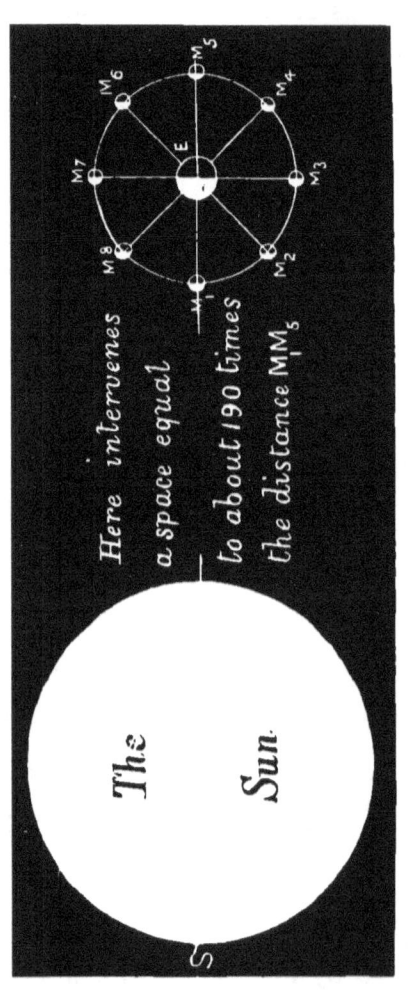

Fig. 12.—Explaining the Moon's changes.

Now, it is clear that when the moon is at M_1, her illuminated face is turned from the earth, E. She therefore cannot be seen; and accordingly, in fig. 2, she is presented as a black disc at 1 to correspond with her invisibility when she is as at M_1. She passes on to M_2; and now from E a part of her illuminated half can be seen towards the sun, which would be towards the right, if we imagine an eye at E looking towards M_2. Her appearance then is as shown at 2, fig. 13. In any intermediate portion between M_1 and M_2, the sickle of light is visible but narrower. We see also that all this time the moon's place on the sky cannot be far from the sun's place, for the line from E to M_2 is not greatly inclined to the line from E to S. When the moon has got round to M_3, the observer on the earth sees as much of the dark half as of the bright half of the moon, the bright half being seen, of course, towards the sun. Thus the moon appears as at 3, fig. 13. Again as to position, the moon is now a quarter of a circuit of the heavens from the sun, for the line from E to M_3 is square to the line from E to S. We see similarly that when at M_4 the moon appears as shown at 4, fig 13, for now the observer at E sees as small a part of the moon's dark side as he had seen of her bright side when she was at M_2. When she is at M_5 the observer at E sees her bright face only, the dark face being turned directly from him. She, therefore, appears as at 5, fig. 13. Also being now exactly

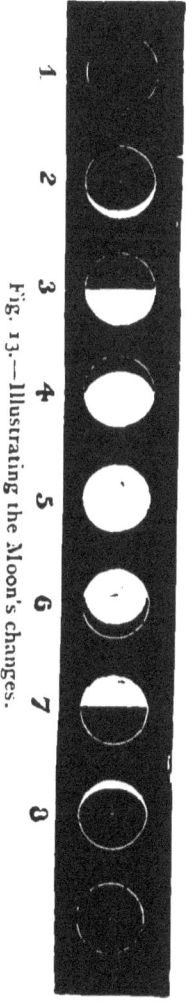

Fig. 13.—Illustrating the Moon's changes.

opposite the sun, as we see from fig. 12, she is at her highest when the sun is at his lowest, or at midnight; and at this time she rules the night as the sun rules the day.* As the moon passes on to M_6, a portion of her dark half comes into view, the bright side being now towards the left, as we look at M_5 from E, fig. 12. Her appearance, therefore, is as shown at 5. When at M_7 she is seen as at 7, half-bright and half-dark, as when she was at M_3, but the halves interchanged. At M_8 she appears as at 8, and, lastly, at M_1 she is again undiscernible.

The ancient Chaldean astronomers could have little doubt as to the validity of this explanation. In fact,

---

* It has been thought by some that, in the beginning, the moon was always opposite the sun, thus always ruling the night. Milton thus understood the account given in the first book of Genesis. For he says,—

> Less bright the morn,
> But opposite in levell'd west was set
> His mirror, with full face, borrowing her light
> From him; for other light she needed none
> In that aspect; and still that distance keeps
> Till night, then in the east her turn she shines,
> Revolv'd on Heav'n's great axle.

It was only as a consequence of Adam's transgression that he conceives the angels sought to punish the human race by altering the movements of the celestial bodies—

> To the blank moon
> Her office they prescribe—

It is hardly necessary to say, perhaps, that this interpretation is not scientifically admissible.

while it is the explanation obviously suggested by observed facts, one cannot see how any other could have occurred to them.

But if they had had any doubts for a while, the occurrence of eclipses would soon have removed those doubts. They must early have noticed that at times the full moon became first partly obscured, then either wholly disappeared or changed in colour to a deep coppery red, and after a while reappeared. Sometimes the darkening was less complete, so that at the time of greatest darkness a portion of the moon seemed eaten out, though not by a well defined or black shadow. These phenomena, they would find, occurred only at the time of full moon. And if they were closely observant, they would find that these eclipses of the moon only occurred when the full moon was on or near the great circle round the stellar heavens, which they had learned to be the sun's track. They could hardly fail to infer that these darkenings of the moon were caused by the earth's shadow, near which the moon must always pass when she is full, and through which she must sometimes pass more or less fully; in fact, whenever, at the time of full, she is on or near the plane in which the earth travels round the sun. Solar eclipses would probably be observed later. For though a total eclipse of the sun is a much more striking phenomenon than a total eclipse of the moon, yet the latter are far more common. A partial eclipse of the sun may readily pass unnoticed, unless the sun's rays are so mitigated by haze or mist that it is possible to look at his disc without pain. Whenever solar eclipses came to be noted, and we know from the Chaldean discovery of the great eclipse period, called the *Saros*, that they were observed at least two thousand years before the Christian era, the fact that the moon is an opaque body circling round the earth, and much nearer to the earth than the sun is, must be regarded as demonstrated. Not only would eclipses of the sun be observed to occur only when the moon was passing between the earth and the sun, but in an eclipse of the sun, whether total or partial, the round black body cutting off the sun's light wholly or partially would be seen to have the familiar dimensions of the lunar orb.

Leaving solar and lunar eclipses for description on another occasion, I will now proceed to consider a peculiarity of moonlight which must very early have attracted attention,—I mean the phenomenon called the harvest-moon.

The moon circuits the heavens in a path but slightly inclined to that of the sun, called the ecliptic, and for our present purpose we may speak of the moon as travelling in the ecliptic. Now we know that during the winter half of the year the sun is south of the equator, the circle of the heavenly sphere which passes through the east and west points of the horizon, and has its plane square to the polar axis of the heavens. During the other or summer half of the year he is north of the equator. In the former case the sun is above the horizon less than half the twenty-four hours, day being so much shorter as the sun is farther south of the equator; whereas in the latter case the sun is above the horizon more than twelve hours, day being so much the longer as the sun is farther north of the equator. Precisely similar changes affect the moon, only, instead of taking place in a year (the time in which the sun circuits the stellar heavens), they occur in what is called a sidereal month, the time in which the moon completes her circuit of the stellar heavens. For about a fortnight the moon is above the horizon longer than she is below the horizon, while during the next fortnight she is below the horizon longer than she is above the horizon. Now clearly when the length of what we may call the moon's diurnal path (meaning her path above the horizon) is lengthening most, the time of her rising on successive nights must change least. She comes to the south later and later each successive night by about $50\frac{1}{2}$ minutes, because she is always travelling towards the east at such a rate as to complete one circuit in about four weeks; and losing thus one day in 28, she losses about $50\frac{1}{2}$

minutes per day. If the interval between her rising and her arriving to the south were always the same, she would rise 50½ minutes later night after night. But if the interval is lengthening, say by 10 minutes per night, she would of course rise only 40½ minutes later; if the interval is lengthening 20 minutes per night, she would rise only 30½ minutes later, and so forth. But the lunar diurnal arc *is* lengthening all the time she is passing from her position farthest south of the equator to her position farthest north, just in the same way as the solar day is lengthening from mid-winter to midsummer, only to a much greater degree. And as the solar day lengthens fastest at spring when the sun crosses the equator from south to north, so the time the moon is above the horizon lengthens most, day by day, when the moon is crossing the equator from south to north. It lengthens, *then*, from an hour to an hour and 20 minutes in one day, that is, the interval between moon-rise and moon-setting increases from 30 to 40 minutes. At this time, then, whenever it happens in each lunar month, the moon's time of rising changes least: instead of the moon rising night after night 50½ minutes later, the actual difference varies only from 10 to 20 minutes.

Now if this happens at a time when the moon is not nearly full, it is not specially noticed, because the moon's light is not then specially useful. But if it happens when the moon is nearly full, it is noticed, because her light is then so useful. A moon nearly full, afterwards quite full, and then for a day or two still nearly full, rising night after night at nearly the same time, remaining also night after night longer above the horizon, manifestly serves man for the time being in the most convenient way possible. But it is clear that as the full moon is opposite the sun, and as to fulfil the condition described we have seen that she must be crossing the equator from south to north, the sun, opposite to her, must be at the part of his path where he crosses the equator from north to south. In other words, the time of year must be the autumnal equinox. Thus the moon which comes to "full" nearest to September 22 or 23 will behave in the convenient way described. At this time, moreover, when she rises night after night nearly at the same time, the nights are lengthening fastest while the time the moon is above the horizon is lengthening still more, and therefore, in all respects, the moon is then doing her best, so to speak, to illuminate the nights. At this season the moon is called the harvest-moon, from the assistance she sometimes renders to harvesters.

The moon which is full nearest to September 22—23 may precede or follow that date. In the former case only can it properly be called a harvest-moon. In the latter it is sometimes called the hunter's moon. The full moon occurring nearest to harvest time will always partake more or less of the qualities of a full moon occurring at the autumnal equinox: and similarly of a full moon following the autumnal equinox. So that, in almost every year, there may be said to be a harvest-moon and a hunter's moon. But, of course, it will very often happen that in any particular agricultural district the harvest has to be gathered in during the wrong half of the lunar month, that is, during the last and first, instead of the second and third quarters.

The reader must not fall into the mistake of supposing, as I have seen sometimes stated in text-books of astronomy, that we are more favoured in this respect than the inhabitants of the southern hemisphere. It is quite true that the same full moon shines on us as on our friends in New Zealand, Australia, and Cape Colony, and also that our autumn is their spring, and their spring our autumn. But the full moon we have in autumn behaves in the southern hemisphere not as with us, but as our spring full moon behaves; and the full moon of our spring, which is their autumn, behaves with them as our autumn moon behaves with us. It is, therefore, for them a harvest-moon if it occur before the equinox, and a hunter's moon if it occur after the equinox. A very little consideration will show why this is. In fact if, in the explanation given above, the words north and south be interchanged, and March 21—22 written for September 22—23, the explanation will be precisely that which I should have given respecting the harvest (or March) moon of the southern hemisphere, if I had been writing for southern readers.

Having thus considered the moon as a light-giver, both in respect of her monthly changes and of that yearly change which causes her services to be most useful in harvest time, let us consider what science tells us of the orb which thus usefully reflects to us the solar rays.

The moon is a globe about $2159\frac{1}{2}$ miles in diameter, travelling round the earth at a mean distance of 238,818 miles. Her path round the earth is not, however, a circle, but an ellipse, which itself is constantly varying in shape. The average eccentricity of the moon's path is such that her greatest and least distances, as she circuits round it, are 251,953 miles and 225,683 miles respectively; but when it is most eccentric, her greatest and least distances are 252,948 miles and 221,593 miles respectively; while, when it is least eccentric, they are respectively 250,324 miles and 227,312 miles. The earth's surface exceeds the moon's nearly $13\frac{1}{2}$ times, the actual number of square miles in the moon's surface amounting to 14,600,000. This is nearly equal to Europe and Africa together, or, more nearly still, to North and South America together, without their islands. In volume our earth exceeds the moon rather more than $49\frac{1}{4}$ times: or, more nearly, if the earth's volume be represented by 10,000, the moon's will be represented by 209. The materials of the moon's globe are either lighter or (more probably) they are less closely compacted than those forming our earth,— for, according to the best modern estimates, the earth exceeds the moon in mass nearly $81\frac{1}{2}$ times. Assuming as the most probable value of the earth's mean density about $5\frac{7}{10}$ times the density of water, the moon's mean density is equal to $3\frac{40}{100}$ times that of water. Gravity at her surface is accordingly much less than at the surface of the earth; a quantity of matter weighing six pounds at the surface of the earth would weigh almost exactly one pound at the surface of the moon.

The moon circuits once round the earth in 27d. 7h. 43m. 11.5s. This is the time in which, viewed from the earth, she seems to complete one circuit round the stellar heavens, and is therefore called a sidereal month. But as the earth is all the time travelling the same way round the sun, the lunar month is longer. Thus, suppose S (fig. 14) to be the sun, E the earth at the beginning of a lunar month, M_1 M_2 M_3 M_4 the moon's

path, and M_1 the moon's place on the line joining E
and S. If the earth remained at rest while the moon
went round the path $M_1 M_3$, then after completing one
circuit the moon would again be at M_1 on the line
joining E and S, or it would be new moon again. But
the earth is moving onwards along the arc EE' of her
circuit round the sun. So that when the moon has

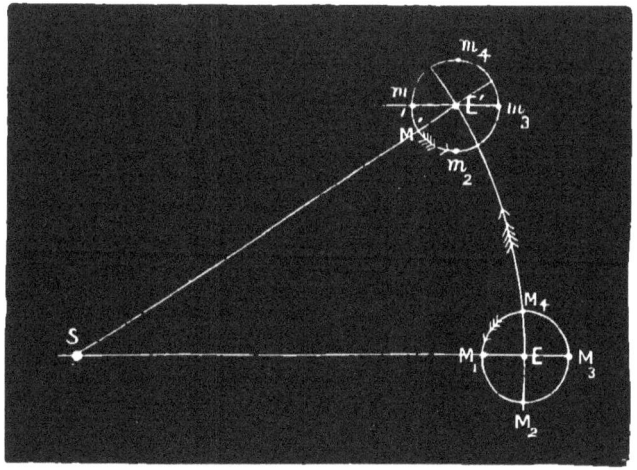

Fig. 14.—Explaining the difference between a sidereal lunar month.
and a common lunar month or lunation.

completed one circuit she is at M_4 (E'm_1 drawn parallel
to EM_1) and has still to travel some distance before she
gets round to M' on the line joining S and E'. The
lunation, or interval between successive new moons, has
an average duration of 29d. 12h. 44m. 38s., exceeding
a sidereal month by 2d. 5h.

It would not, however, be correct to regard the earth

as the true centre of the moon's motion. The moon is in reality a planet circling round the sun, but largely perturbed by the attraction of its companion planet the earth. If the moon's path in the course of a year were carefully drawn to scale, or, better, were modelled by means of a fine wire, it would scarcely be distinguishable from a similar picture or model of the earth's path round the sun. Or thus, the entire width of the moon's track is about 477,636 miles, while the diameter of the orbit along which she and the earth both travel is nearly 104,000,000 miles, or 385 times as great. If we draw then a circle $3\frac{85}{100}$ inches in diameter to represent the earth's path round the sun, somewhat eccentrically placed, and the circular line is 1-100th of an inch wide, the moon's track would be fairly represented by a curve touching alternately the inside and the outside edge of this circular line, at equidistant points dividing the circle into about $24\frac{3}{4}$ parts.

Regarding the moon as a planet, she may be said to have a year, and seasons, and day and night, as the earth has, but very unlike our seasons and days. Her axis is inclined only $1\frac{1}{2}$ degrees from uprightness to her path, whereas our earth's axis is inclined $23\frac{1}{2}$ degrees. The sun's range of mid-day altitude is in fact not quite equal to the range of our sun in mid-day height, from four days before to four days after either spring or autumn. The lunar day lasts a lunar month, daytime and night-time each lasting rather more than a fortnight. The lunar year of seasons is not, as is commonly stated, the same in length as ours. She goes round the sun in the same time, so that her sidereal year is the same as ours; but owing to the swaying round of her axis her year of seasons or tropical year is shorter. Our tropical year is also shorter than the sidereal year, but very little shorter, because the earth's axis sways round once only in 25,868 years. The moon's axis sways round once in $18\frac{2}{5}$ years, and accordingly the year of seasons is much more effectively shortened. It lasts, in fact, only 346d. 14h. 34m. of our time; and contains only $11\frac{3}{4}$ lunar days. So that I cannot altogether agree with Sir W. Herschel's statement, that "the moon's situation with respect to the sun is much like that of our earth, and by a rotation on its axis it enjoys an agreeable variety of seasons, and of day and night."

When the moon is examined with a telescope her surface is seen to be marked by many irregularities. There are large dark regions which were formerly thought to be seas, but are now know to be land-surfaces. Some of these regions are singularly level, and have been thought to be old sea-bottoms. Mountains and mountain ranges are another important feature of the moon's surface. Some, like our Rocky Mountains and Andes, form long continuous chains; others form elevated plateaus whence ridges extend in various directions. A very striking form is that of narrow ridges little raised above the general level, but reaching over enormous areas of the moon's globe. It is a system of this kind, radiating from a great lunar crater called Tycho, which gives to small photographs of the moon the appearance of a peeled orange. They are supposed to indicate the action of tremendous forces of upheaval, in past ages, bursting open portions of the moon's crust.

But the most characteristic of all the lunar features are the crater mountains, which exist on a scale not only much larger relatively to the moon's globe than the scale on which terrestrial craters are formed, but much larger absolutely. They are also far more numerous. Some parts of the moon's surface, especially in the bright south-western quarter of her face, are literally crowded with craters of various dimensions.

There are few signs of the former emission of lava from the lunar craters. Within some of them recent changes have been suspected. A remarkable instance is that of the crater Linné, marked in Mädler's map as a deep, well-walled crater, some four miles in diameter. At present only a small crater can be seen in its place. The surrounding region is rather conspicuously bright.

It is not necessary to infer that there has been any volcanic disturbance, however. Far more probably the walls have been thrown down through the long-continued action of that alternate expansion and contraction, which must affect the moon's crust as the long fortnightly day proceeds, and then the equally long lunar night.

There are many well-marked valleys on the moon, besides clefts and ravines. The features called *rilles* are among the most perplexing objects on the moon's surface. Webb, in his charming and most useful little book, "Celestial Objects for Common Telescopes," thus describes them: "These most singular furrows pass chiefly through levels, intersect craters (proving a more recent date), reappear beyond obstructing mountains, as though carried through by a tunnel, and commence and terminate with little reference to any conspicuous feature of the neighbourhood. The idea of artificial formation is negatived by their magnitude; they have been more probably referred to cracks in a shrinking surface." Some observations would seem to show that they have been formed from rows of closely-adjacent small craters. *Faults*, also, or closed cracks where the surface is higher on one side than on the other, have been recognised from the careful study of the shadows on the moon's disc.

From measurements of the shadows of lunar mountains, it appears that their average height is about five miles. In comparing this elevation with that assigned to terrestrial mountains, it must be remembered that these are measured from the sea-level; if the average height of terrestrial mountains were determined with reference to the sea-bottom it would be far greater. Still, even taking this circumstance into account, the average height of the lunar mountains bears a far greater ratio to the diameter of the globe on which they stand than the average height of our mountains to the earth's diameter.

Several circumstances agree in showing that the moon's atmosphere must be exceedingly rare. The shadows of lunar mountains are either actually black or nearly so. When the moon hides the sun in total eclipse, no sign can be seen of any refractive effort exerted on the sun's rays. When a star is hidden (or *occulted*) by the moon, the star vanishes in an instant and reappears with equal suddenness. It is certain from these phenomena that the moon has either no air, or air exceedingly tenuous. It is equally clear that she has no water, for if she had we should undoubtedly be able to recognise the occasional formation or dissipation of mist and vapour over parts of the moon's surface. No signs of such phenomena have ever been observed. The moon is certainly at present a waterless globe, so far at least as her surface is concerned.

It has been thought that though there is no water and

very little air on the side of the moon turned towards the earth, there may be both water and air on the farther unseen side. The theory has been long since given up, but the reasoning on which it depends is worth noting. Owing to the strange circumstance that the moon rotates on her axis in the same time in which she revolves round the earth, she always presents the same face towards the earth, or very nearly so. If her axis were exactly square to the path in which she circuits the earth, and if she revolved at a uniform rate, we should have exactly the same side constantly turned towards us. But as the axis is inclined about $6\frac{2}{3}°$ from uprightness to the path round the *earth* (which, be it remembered, is not in the same plane as the path round the sun, but inclined 5° 8′ to it), the northern and southern parts of the moon are alternately swayed over by about $6\frac{2}{3}°$ into view. This apparent swaying is called a libration, and the libration just described is called the libration in latitude. Again, as the moon does not travel at a uniform rate round the earth, but faster than her mean rate when nearer to us, and slower when farther from us, she alternately gains and loses in her motion of revolution as compared with her motion of rotation, by a quantity varying between 5° and $7\frac{3}{4}°$, to which varying extent the parts east and west of her mean disc are alternately swayed into view. This is called the libration in longitude. Thus we see, beyond the edge of the *mean* half turned towards us, a considerable fringe of the other half. If a globe, as PAP'B, fig. 15, were divided into two

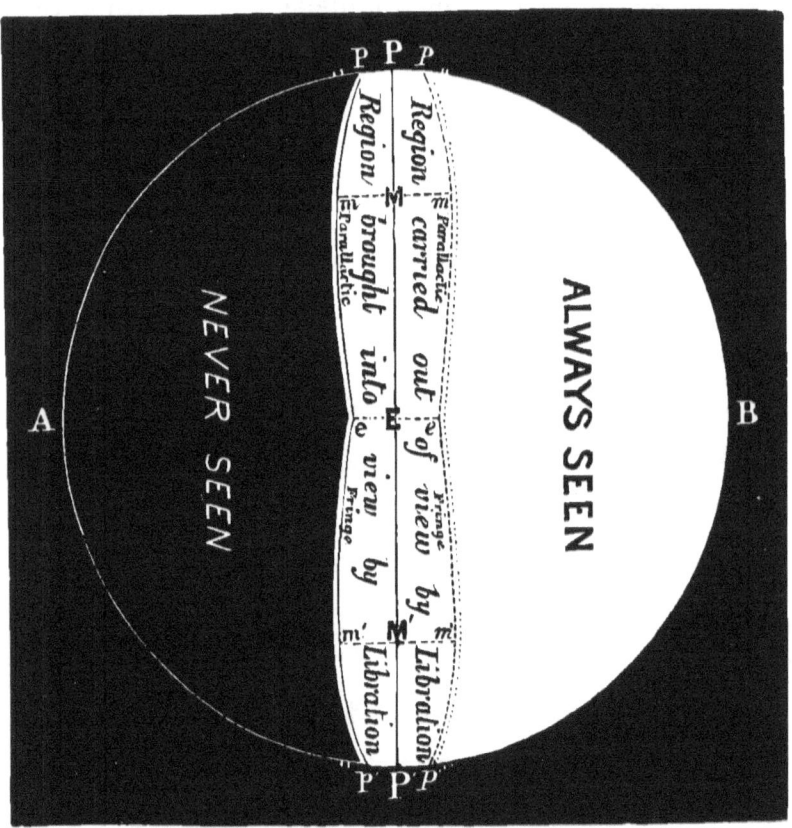

Fig. 15.—Illustrating lunar libration.

halves to represent the farther and nearer halves of the moon, and held so that that dividing circle were seen as PEP' in the figure, then Ppep'P' would represent the part brought into view at different times by the apparent swaying described above; while P*pep*'P' would represent the parts swayed out of view. The regions thus alternately in view and out of view have their greatest breadth, not at the poles or east and west, but at mMm and m'M'm', where the two librations act together. The narrow fringe bordering these regions is that brought into or out of view by changes in the place of the observer on earth, due to the earth's rotation. It is called the parallactic fringe, any change in the apparent position of a heavenly body, or part of one, on account of the earth's rotation, being termed *parallax*.

Lastly, let us return to the consideration of moonlight, as depending on the condition of the moon's surface. To one who observes the moon as seen on the sky, her light appears white; but it must not be supposed that she is a white body. Careful estimates of the quantity of light she reflects show that she is more nearly black than white, though in reality she is neither one nor the other. It has been said, and truly, that if the surface of the moon were covered with black velvet she would still appear white; for even black velvet reflects some light, and whatever light the moon reflected would show her by contrast with the blackness of the sky, as a luminous body or white. It follows from the observations made by Zöllner that if the moon's surface were covered with white snow she could give us about $4\frac{1}{2}$ times as much light as she actually does. If she were covered with white paper she would give more than 4 times as much light as she does. If she had a surface of white sandstone her light would be nearly half as great again as it is. She gives rather more light than she would if her surface consisted entirely of weathered grey sandstone, or of clay marl, and more than twice as much light as she would give if her surface were of moist earth, or dark grey syenite. As some parts of her surface are obviously much brighter than others, we must infer that some parts shine with much more, and others with much less, brightness than weathered grey sandstone. Probably some parts are much brighter than white sandstone, and some much darker than dark grey syenite. From the degree in which her lustre changes with her changing aspect, Zöllner infers that her mountains have an average slope of about fifty-two degrees.

IX

THE PLANET MARS.

EVERY one who notices the stars at all,— and who that thinks and can see does not? —must have observed during the autumn of 1877 two bright stars in the southern heavens. One of these shone with a lustre which but for its ruddy hue would have caused the star to be taken for the planet Jupiter; the other shone with a somewhat yellowish light, and was much fainter, though surpassing most of the *fixed* stars in brightness. The former was the planet Mars, the latter the ringed planet Saturn. The motions of these two stars with respect to each other and to the neighbouring stars were sufficiently conspicuous to attract attention. During October these stars attracted still more attention, because they drew nearer and nearer together, to all appearance, until on November 4th they were at their nearest, when the distance separating them was about one-third the apparent diameter of the moon, so that in a telescope showing at one view the whole disc of the moon, Mars and Saturn on the night of November 4th appeared like a splendid double star, the primary a fine red orb, the companion a smaller body, but attended by a splendid ring system and companion moons.

It was strange when we looked at these two stars, the yellow one apparently much smaller than the brighter, and the pair seemingly close together, to consider how thoroughly the reality differed from these appearances. The fainter and seemingly the smaller of the two stars was in reality some four thousand times larger than the brighter, and had, among eight orbs attending upon it, one nearly as large as the ruddy planet which as actually seen so completely outshone Saturn himself. Again, instead of being near each other, those two bodies were in reality separated by a distance exceeding some sixteen times that which separated us from the nearer of the two.

I propose now to consider some of the more interesting characteristics of these two planets, presenting specially those features which mark Saturn as the representative of one family of bodies, and Mars as the representative of another and an entirely different family.

It will be well to consider Mars first; for although, as will presently be seen, Saturn came earlier of the two

Fig. 16.—The paths of Mars and Saturn during the autumn of 1877.

to the portion of his path where he was most favourably seen, there was nothing specially remarkable about the approach of Saturn on that occasion, whereas Mars in the year 1877 made a nearer approach to the earth than he has for thirty-two years past, or will for some forty-seven years to come.

In the first place, let us note the apparent paths on which the two planets have been and are now travelling.

Fig. 16 presents that part of the zodiac along which lay the apparent paths of Mars and Saturn in 1877. The stars marked with Greek letters belong to the constellation Aquarius, or the Water-Bearer (his jar is formed by the stars in the upper right-hand corner of the picture),—with a single exception, the star marked κ, which, with those close to it not lettered, belongs to the constellation Pisces, or the Fishes. Thus the loops traversed by the two planets in 1877 both fell in the constellation of the Water-Bearer; but, as will be seen from the symbols on the ecliptic, these loops lie in the zodiacal sign Pisces, which begins at κ and ends at ♈. The signs have long since passed away, in fact, from the constellations to which they originally belonged.

It will be noticed that Mars described a wide loop ranging to a considerable distance from the ecliptic (or sun's track). Saturn, on the other hand, travelled on a narrow and shorter loop lying much nearer to the ecliptic, his whole track, except just where he was turning,—his stationary points,—lying nearly parallel to the ecliptic. It may be well to mention the reason of this well-marked difference. Mars does not in reality range even quite so widely from the plane of the ecliptic as Saturn does. Nay, his path is even less inclined to the ecliptic. (This may sound like repetition, but the inclination of a planet's path to the ecliptic is one thing, the range of the planet north and south of the ecliptic, in miles, is another. Mercury, for example, has of all planets the path most inclined to the ecliptic, but Mercury never attains anything like the same distance from the plane of the ecliptic which is attained by the remote planet Uranus, whose path is of all others the least inclined to the plane of the ecliptic. In fact, none of the planets, except Venus and Mars, have so small a range from the ecliptic in actual distance as Mercury has.) The reason why the range of Mars from the ecliptic appeared so much greater than that of Saturn, in 1877, is similar to the reason why Mars, though much smaller than Saturn, largely outshone him. Mars looked larger because he was nearer, his loop looked larger because his real path was nearer. For the same reason that a hut close by seems to stand higher above the horizon than a palace at a distance, or a mountain yet further away, so the displacement of Mars from the ecliptic plane appeared greater than that of Saturn, though in reality much less.

Let us consider how the paths of these planets are really situated. I know of no better way of showing this than by drawing the paths of the two families of planets separately. It is in fact utterly impossible to give an accurate yet clear view of the solar system in a single picture; and the student may take it for granted that every drawing or plate in which this has ever been attempted is from one cause or another misleading.

In figs. 17 and 18 the shape and position of the planetary paths are correctly shown. Very little description is necessary, but it may be mentioned that on each orbit the point nearest to the sun is indicated by the initial letter of the planet, while the point farthest from the sun is indicated by the same letter *accented*. The places where each path crosses the plane of the earth's—which is supposed to be the plane of the paper—are marked ☊ and ☋, the former sign marking where the planet in travelling round in the direction shown by the arrows crosses the plane of the earth's path from below upwards, while the latter marks the place where the planet in travelling round crosses the plane of the earth's path from above downwards.

Fig. 17 shows the paths of the inner family of planets of which our earth is a member. Fig. 18 shows the outer

156   *THE PLANET MARS.*

family of planets, and inside of it the ring of small planets called asteroids. Inside that ring, again, we see the paths of the inner family of planets; but they

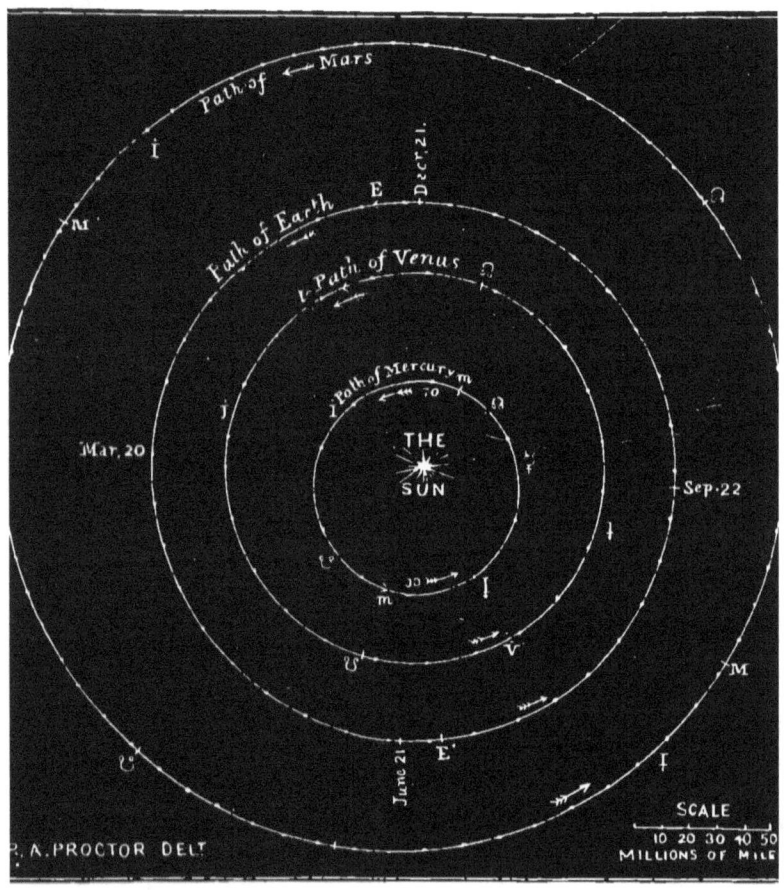

Fig. 17.— The paths of Mercury, Venus, the Earth, and Mars, around the Sun.

appear on a very small scale indeed. In fact, the scales appended to the two figures show that a length which represents 50,000,000 miles in fig. 17, represents

1,000,000,000 miles in fig. 18; or, in other words, the scale of fig. 18 is only one-twentieth of the scale of fig. 17. On the scale of fig. 17 the sun would be fairly

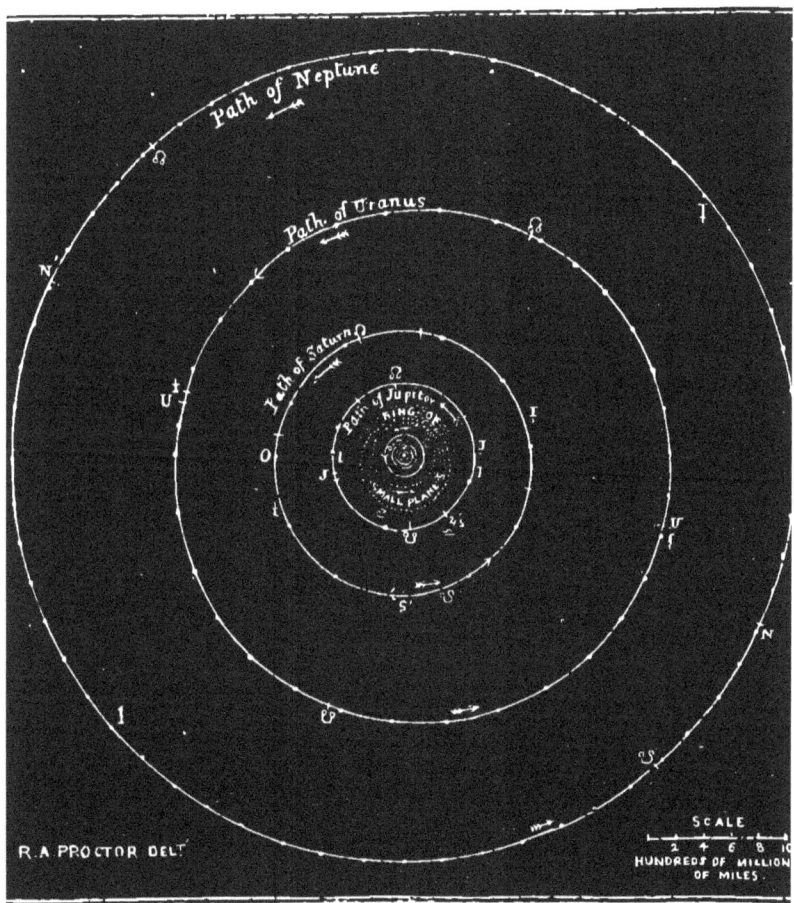

Fig. 18.—The paths of Jupiter, Saturn, Uranus, and Neptune, around the ring of small planets.

represented by an ordinary pin-hole; on the scale of fig. 18 the sun would be scarcely visible. The dots round the orbits show the planets' places at intervals of 10 days in fig. 17, and of 1000 days in fig. 18, starting always from the left side of orbit (on horizontal line through sun).

Now looking at fig. 18 and noting how small is the distance of the path of Mars from the earth's path, compared with the distance of Saturn's path, we understand why Saturn, despite his far superior size, shines far less brightly in our skies than Mars does. In fact, in October, 1877, the Earth and Mars were on the parts of their tracks which lay nearest together, that is, the parts occupying the lower right-hand corner of fig. 17; and turning to fig. 18, we perceive that the distance separating the two paths here is very small indeed compared with Saturn's distance.

So that, when we looked at Mars and Saturn as they shone in conjoined splendour in our skies, in 1877, we saw in the bright orb of Mars the planet whose track lies nearest to us in that direction, whereas in looking at Saturn the range of view passed athwart the track of Mars, through the ring of asteroids, and past the orbit of Jupiter, before entering the wide and barren region which separates the orbits of the two giant members of the solar system.

We study Mars under much more favourable conditions than either Jupiter or Saturn. And yet, at a first view, the telescopic aspect of this interesting planet is exceedingly disappointing. Galileo, who quite easily discovered the moons of Jupiter with his largest telescope, could barely detect with it the fact that Mars is not quite round at all times, but is seen sometimes in the shape of the moon two or three days before or after full. "I dare not affirm," he wrote on December 30, 1610, to his friend Castelli, "that I can observe the phases of Mars; yet, unless I mistake, I think I already perceive that he is not perfectly round." But even in a large telescope one can see very little except under very favourable conditions. It has only been by long and careful study, and piecing together the information obtained at various times, that astronomers have obtained a knowledge of the facts which appear in our text-books of astronomy. The possessor of a telescope who should expect, on turning the instrument towards Mars, to perceive what he has read in descriptions of the planet, would be considerably disappointed.

First noticed among the features of the planet were two white spots of light occupying the northern and southern parts of his disc. These are now known to be regions of snow and ice, like those which surround the poles of our own earth. But how different the reality must be from what we seem to see in the telescope! These two tiny white specks represent hundreds of thousands of square miles covered over with great masses of snow and ice, which doubtless are moved by disturbing forces similar to those which make our arctic regions for the most part impassable even for the most daring of our seamen.

The snow-caps of Mars change in size as the planet circuits round the sun, completing his year of seasons (which lasts 687 of our days). They are largest in the winter of Mars, smallest in the Martian summer; so that, as it is winter for one hemisphere when it is summer for the other, one of the snow-caps is larger than the other at the winter and summer seasons. In the same way, our arctic snows extend more widely during our winter, while the antarctic snows then retreat; whereas, during our summer, when it is winter in the southern hemisphere, the antarctic snows advance and our arctic snows retreat.

But we have still to learn why these white spots are *known* to be masses of snow. They might well from analogy be considered to be snows, since they behave like the snows of our polar regions. Yet that would be very different from proving them to be snow masses. I shall now show how this has been done, and afterwards describe the lands and seas of the planet, and give a short account of the recent interesting discovery of two moons attending on the planet which Tennyson had called the "moonless Mars."

Even before the poles of Mars had been discovered,

observers had perceived that the planet has marks upon its surface. Cassini, in 1666, at Paris, found by observing these spots that the planet turns on its axis once in about twenty-four hours forty minutes. In the same year Dr. Hooke observed Mars. He was in doubt whether the planet turned once round or twice round in about twenty-four hours; for with his imperfect telescope two opposite faces of the planet seemed so much alike that he was doubtful whether they really were two different faces or the same. Fortunately he published two pictures of the planet, taken on the same night in March, 1666, and we have been able to keep such good count of Mars's turning on his axis, that we know exactly how many times he has turned since that distant time. However, at present, we need not further consider the turning motion of Mars, but rather what the telescope has shown us about him. Only, let it be remembered that he has a day of about twenty-four hours thirty-seven minutes, and is in this respect much like our earth.

Maraldi, Cassini's nephew, early in the last century observed several spots on Mars, and, in particular, one somewhat triangular dark spot, which was one of Hooke's markings, but more clearly seen by Maraldi. About this time it was seen that the darker markings have a somewhat greenish colour; and towards the end of last century, or, more exactly, about a hundred years ago, the idea was maintained by Sir W. Herschel that the dark-greenish markings are seas, while the lighter parts of Mars, to which the planet owes its somewhat ruddy colour, are lands. Sir W. Herschel also was the first to show that Mars, like our earth, has seasons. It had been supposed by Cassini, Maraldi, and others, that the axis of Mars is upright to the level of the path in which he travels. Of course, if this were so, the light of the sun would always fall on the planet in the same way; for the sun is in that level. But the axis, like that of our own earth, is bowed considerably from uprightness; so that at one part of his year the sun's rays fall more fully on his northern regions, and his southern regions are correspondingly turned away from the sun; then it is summer in his northern regions, winter in his southern. At the opposite season the reverse holds, and then winter prevails over his northern and summer over his southern regions. Midway between these two seasons, the sun's rays are equably distributed over both hemispheres of Mars, and then the days and nights are equal, and it is spring in that hemisphere which is passing from winter to summer, and autumn in the other hemisphere which is passing from summer to winter. All these changes are precisely like those which take place in the case of our own earth. Only, the year of Mars, and therefore his seasons, are longer. He takes 687 days in travelling

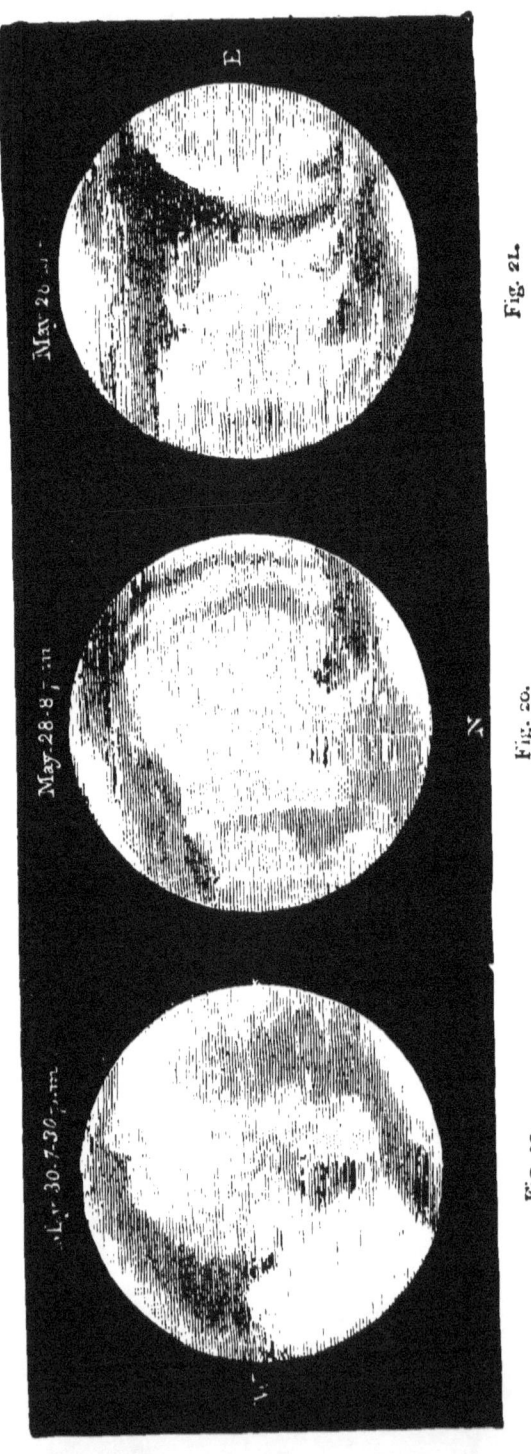

Figs. 19—21.—Three Views of Mars.

round the sun, giving nearly 172 days, or more than five and a half of our months, for each season.

Figs. 19, 20, and 21 are three views of Mars, drawn by Mr. Nathaniel Green, an excellent observer, who has paid special attention to this planet. Fig. 19 shows a faintly-marked sea running north and south (the upper part of the picture being the south, because that is the way in which the telescope used by astronomers inverts objects.) This is one of the markings which deceived Hooke. This picture was drawn on May 30, 1873, at half-past seven in the evening. The second picture was drawn two days earlier, at eight in the evening; but it shows the planet as it would have looked on May 30 at about a quarter past nine in the evening, by which time the sea running north and south had been carried over to the right and lost to view. But another north and south sea had come into view on the right. The third picture shows a view taken three hours later, or at eleven on May 28, when the planet appeared precisely as he would have appeared at a quarter past eleven in the early morning of May 31, had weather then permitted Mr. Green to continue his observations. You see in it the great north and south sea which Maraldi had noticed, the other of those two which had deceived Hooke.

It will be seen from these drawings, which, be it remembered, were taken at the telescope, that it is possible from a great number of such drawings to make a chart of Mars, showing its lands and seas not as they are seen in the telescope, but as they might be laid down by inhabitants of Mars in a map or planisphere. This has been done, with gradually increasing accuracy,—first by Sir W. Herschel, next by Beer and Mädler, then by Phillips, and lastly by myself. (In claiming for my own chart greater accuracy, I am simply asserting the superior completeness of the list of telescopic drawings which I was able to consult.) The result is shown in the accompanying chart (fig. 22), which presents the whole surface of Mars divided into lands and seas and polar snows, with the names attached of various observers who have at sundry times contributed to our knowledge of the planet's features.

But now it will be asked by the thoughtful reader, how can any one possibly be sure that the regions called continents and seas do really consist of land and water? At any rate, the doubt might well be entertained respecting the water. For land is a wide term, including all kinds of rock surface, sand, earthy soil, and so forth; but it may seem to require proof that the substance we call water really exists out yonder in space, either in the form of snow and ice at the Martian poles, or as flowing water in the Martian seas, or in the vaporous form in the planet's air.

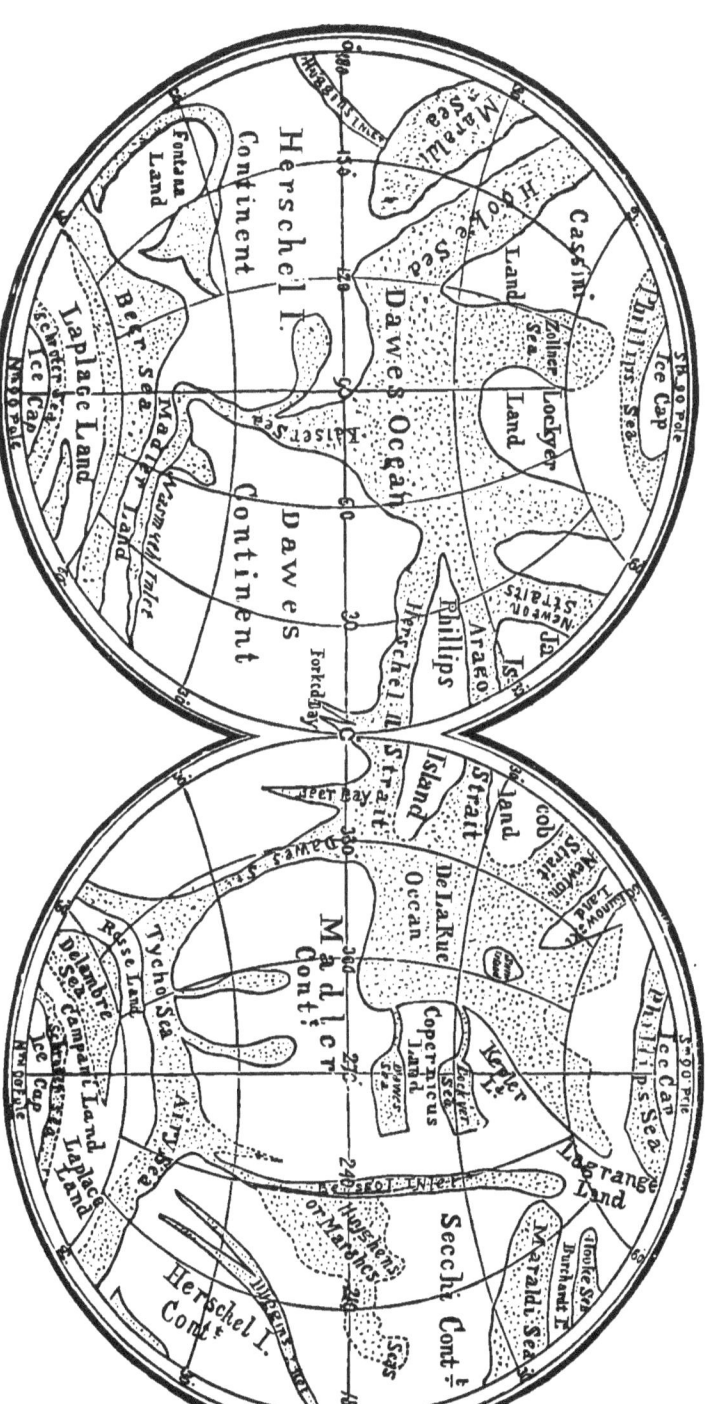

Fig. 22.—Chart of Mars, from 27 drawings by Mr. Dawes.

Very strange, then, at first must the statement seem, that we are as sure of the existence of water in all these forms on Mars as if we had sent some messenger to the planet who had brought back for study by our chemists a block of Martian ice, a vessel full of Martian water, and a flask of Martian air saturated with aqueous vapour. Indeed, I do not know of any discovery effected by man which more strikingly displays the power of human ingenuity in mastering difficulties which, at a first view, seem altogether insuperable. When we know that a mass of ice as large as Great Britain would appear at the distance of Mars a mere bright point; that a sea as large as the Mediterranean would appear like a faint, greenish-blue, streak; and that cloud masses such as would cover the whole of Europe would only present the appearance of a whitish glare, how hopeless seems the task of attempting to determine what is the real chemical constitution of objects thus seen! It might well be thought that no possible explanation of the method used by astronomers could serve to establish its validity. Yet nothing can be simpler than the principle of the method, or more satisfactory than its application in this special case.

First, let the reader rid his mind of the difficulty arising from the enormous distance of the celestial bodies. To do this let him note that there are some things which a body close by can tell us no more certainly than a remote body. For instance, we are just as certain that Mars is a body capable of reflecting sunlight as we are that a cricket-ball is. We know as certainly, too, that the quality of Mars is such that more of the red of the sun's light is sent to us than of the other colours. For we perceive that Mars is a ruddy planet. Since distance in no way interferes with our perception of these general facts, and others like them, we need not necessarily find in mere distance any difficulty in the way of recognising some other facts. All that we require to be shown before admitting the validity of the evidence is, that it is of such a kind that distance does not affect its *quality*, however much distance may and must affect the quantity of evidence.

Now there is a means of taking the light which comes from a body shining either with its own or with reflected light, and analyzing it into its component colours. The spectroscope is the instrument by which this is accomplished. I do not propose to describe here the nature of this instrument, or the details of the various methods in which it is employed. I note only that it separates the rays of different colour coming from an object, and lays them side by side for us,—the red rays by themselves, the orange rays by themselves, and so with the yellow, green, blue, indigo, and violet. And not only are the rays of these colours set by themselves, but the red rays are sorted in order, from the deepest brown-red * to a tint of red (the lightest) which must almost be called orange; the orange in order, from orange which must almost be called red to a tint (the lightest orange) which must almost be called yellow; the yellow, from an almost orange yellow to a yellow just beginning to be tinged with green; the green, from an almost yellow green (the lightest) to a green which may almost be called blue (the darkest); the blue, from this tint to the beginning of the indigo; the indigo, from this tint to the first rays of the violet; and lastly the violet, through all the tints of this beautiful colour to a blackish-brown violet, where the visible spectrum ends. All these tints are sorted in order by the spectroscope, just as a skilful colourist might range in due sequence a myriad tints of colour. But this is only true of really white light, such light as comes from a glowing mass of metal burning at a white heat. In other cases (even when the light may seem white to the eye) some of the tints are found, when the spectroscope spreads out the colours for us, to be missing. And we know that this may be caused in two ways. Either the source of light never gave out those

---

* Brown is not the right word for the tint of red where the visible spectrum begins. I know, however, of no word properly expressing the colour.

missing tints; or, the source of light gave them out, but some absorbing medium stopped them on their way before they reached the spectroscope with which we examine them. There may be cases where we cannot tell very easily which of these is the true cause. But sometimes we can, as the instances I have now to deal with will show you.

The sun's own light shows under this method of spectroscopic analysis millions of tints, in fact I might say millions of red tints, and so forth, right through the spectral list of colours. But also many thousands of tints are wanting. Imagine a rainbow-coloured ribbon, the colours ranged along its length, so that the ribbon is black at both ends, and that from the black of one end the colour merges into very deep red, and thence by insensible gradations through orange, yellow, green, blue, indigo, and violet, into the black of the other end. Then suppose that tens of thousands of the fine threads which run athwart the ribbon—*i.e.*, the short cross threads—are drawn out. Then the ribbon, laid on a dark background showing through the spaces where the threads were drawn out, would represent the solar spectrum. We know then that the light of the sun's glowing mass either wants particular tints originally, or shines through vapours which prevent the free passage of rays of those colours. Both causes might be at work, not one only. At present we are not concerned with this particular point; but I only mention that, in reality, no tints are actually wanting, though some are very much enfeebled.

The sun's light falling on any opaque object is reflected. If the object is white, the light gives exactly the same spectrum, only fainter. Thus, I take a piece of white paper on which the sun's rays are falling, and examine its light with one of Browning's spectroscopes. I get the ordinary solar spectrum. The cold white paper gives me in fact a spectrum which speaks of a heat so intense that the most stubborn metals are not merely melted but vaporized in it. But this heat resides in the sun, not in the paper.

Now, speaking generally, Mars also sends us sunlight, so that when we spread out with the spectroscope the rays coming from this planet, we get the solar spectrum, only of course very much enfeebled. *But* close examination shows that other tints besides those missing from the solar spectrum are missing from the spectrum of Mars. He reflects to us the sunlight, almost as it reaches him, but he abstracts from it a few tints on his own account.

When we inquire what these tints are, we find that they are tints which are *sometimes* wanting even from direct sunlight. When the sun sinks very low and looks like a great red ball through the moisture-laden air, his spectrum is not the same exactly as that of the sun shining high in the mid heaven. It shows other gaps than those corresponding to the ordinary myriads of missing tints. Its red colour shows indeed that something has happened to the sunlight; but, oddly enough (at first sight at least), the gaps are chiefly in the red part of the spectrum, just what one would expect if the sun's light showed a want instead of an excess of ruddy light. The fact is, however, that the violet, indigo, and blue are weakened altogether, not by the mere abstraction of tints here and there. The red suffers under a few abstractions of tint, but remains on the whole little weakened. Now the same gaps which at such times appear in the spectrum of the sun are found (generally, if not always) in the spectrum of the planet Mars, even when he is shining high in the heavens, so that his light is *not* at the time absorbed by the denser portions of our air. In fact the gaps have been seen in the spectrum of Mars when the planet has been shining higher in the heavens than the moon, whose spectrum was found on trial (at the time) not to show the same gaps,—as of course it must have done, and even more markedly, if the missing tints had been abstracted by our own air.

No doubt can remain, then, that the sun's light, which reaches us after falling on Mars, has suffered *at Mars* the same absorption which our own air produces on the rays of the sun when he is low down. But we know what it is in our air which causes this absorption. It is the aqueous vapour. We know this from several independent series of researches. It was proved first by an American physicist, Professor Cooke of Harvard, who found that these lines in the red are always darker when the air is moister. Then by Janssen, who observed the spectrum of great bonfires lit at a distance of many miles, on the Swiss mountains, finding these same lines in the spectrum of the fire-light when the air was heavily laden with moisture. Wherefore we know that the air of Mars must also contain the same substance—the vapour of water—which, in our own air, produces these dark lines. We can, indeed, understand that the ruddy colour of Mars is in part due to this moisture, which, precisely as in our own air it makes the sun and moon look red, would, in the air of a planet, make the planet itself look red.

But how much follows from the discovery that there is moisture in the air of Mars! This moisture can only come from water in sufficient quantities. There must, therefore, be seas on Mars. We should be sure of this from the spectroscopic evidence, even without the evidence given by the telescope. We cannot doubt for a moment, however, knowing as we do how the telescope shows greenish markings on Mars, that these really are the seas and oceans of the planet. And again, the white spots at the poles of Mars can no longer be regarded doubtfully. If we could not see them, but knew only, from the spectroscopic evidence, that Mars must have large seas, we should be sure that his polar regions must be covered with everlasting ice and snow, varying with the seasons, but always surrounding, in enormous masses, the poles themselves. Seeing that the telescope presents spots to our view which, long before the spectroscopic evidence had been obtained or hoped for, had been regarded as analogues of our polar snows, we can now entertain no manner of doubt that they really are so.

But again, recognising the presence of enormous masses of snow and ice around the poles of Mars, and knowing that not only are there wide oceans, seas, and lakes, but that there is an atmosphere capable of carrying mist and cloud, how many circumstances, corresponding to those which we associate with the wants of living creatures, present themselves to our consideration! It remains that I should now consider some of these points.

We have seen that Mars has water in all its forms, solid, liquid, and vaporous. We perceive also that his polar regions do not extend very much farther towards his equator than do the polar ice and snows of our own earth. (Of course the former do not extend so far in actual distance; I refer to their extent compared with the globe they belong to.) It would appear then, at a first view, that the climate of Mars cannot be very unlike that of our earth. Yet this is scarcely possible. For Mars is so much farther than we are from the sun that he receives less than half as much light and heat from that luminary. And it is not easy to conceive that the deficiency can be compensated by any effects due to the nature of the Martian air. It is more likely by far that this air is much rarer than that it is much denser than ours. For not only can it be shown that with the same relative quantity of air a smaller planet would have a smaller quantity above each square mile of its surface than would a larger one,* but the gravity at the surface

---

* Suppose there are two planets A and B of equal density, of which A has a diameter twice as great as that of B. Then the volume of A is eight times greater than B's volume. So that if the volume of its atmosphere exceed the volume of B's air in the same degree, the planet A has eight times as much air as the planet B. But the surface of A is only four times as great as the surface of B; so that if A had only four times as much air as B, there would be the same quantity of air above each square mile of A's surface as above each of B's surface. Since then A has eight times—not merely four times—as much air as B, it follows that A has twice as much air over each square mile of surface as B has. And similarly in all such cases, the general law being that the larger planet has more air over each

of the smaller planet being less, the air there is much less compressed by its own weight (having in fact much less weight), and is therefore rarer. Thus the probability is that the air of Mars is like that at (or even above) the summits of our highest mountains, where we know that an intense cold prevails. It is not that the sun's rays do not fall there with as much heating power as at the sea-level, for experiment shows that they fall with even greater power. But there is less air to be warmed and to retain the heat. The difference may be compared in fact to that between a well-watered country near the sea and an arid desert. The sun's rays fall as fiercely on one as on the other, but because there is no moisture in the desert to receive (after the fashion characteristic of water) the solar heat and retain it, the heat passes away so soon as the sun has set, and intense cold prevails, while over the well-watered region the temperature is much more uniform, and warm nights prevail. So is it at the summits of lofty mountains. The sun's rays are poured on them as hotly as elsewhere, but there is little air to retain the moisture, so that the heat passes away almost as quickly as it is received, and during the night as much fresh snow is formed as had been melted during the day. And so it would certainly be with Mars, if, other things

square mile of surface in the same degree that its diameter exceeds that of the other.

being the same, the air were as rare as it is at the summits of our loftiest mountains. If, as seems probable, the air is still rarer than this, the cold would be still more intense.

It would seem, then, that either some important difference exists, by which the Martian air is enabled to retain the sun's heat even more effectively than our air does (for the climate as indicated by the limits of the polar snows seems the same, though the distance from the sun is greater); or else there is some mistake in the supposition that the same general state of things prevails on Mars as on our own earth.

I confess that though Professor Tyndall has shown clearly how the atmosphere of a more distant planet *might* make up for the deficient supply of solar heat, by more effectively retaining the heat, I know of nothing in either the telescopic or the spectroscopic evidence respecting any of the planets which tends to show, or even renders it likely, that any such arrangement exists,— excepting always the peculiarity in Mars's case which we are now endeavouring to explain. Insomuch that should any other explanation of the difficulty be suggested, and appear to have weight in its favour, I apprehend that the mere possibility of an atmospheric arrangement, such as has been suggested, should not prevent our admitting this other explanation.

I am inclined to think that there is such an explanation. It seems to me that there are good reasons for regarding Mars as a planet which has passed to a much later stage of planetary life than that through which our earth is now passing, and that in this circumstance some of the peculiarities of his appearance find their explanation. As a planet outside the earth, Mars must probably be regarded as one formed somewhat before the earth. As a much smaller planet, he would be not only less heated when first found (whatever theory of planetary formation we adopt), but would also have parted much more rapidly (relatively) with his heat, according to the same law which makes a small mass of metal cool more quickly than a large one. If he has a rarer atmosphere he would be a colder planet on that account also. Being also remoter from the sun, he receives less heat from that orb, and we thus have a fourth reason for regarding Mars as a much colder planet than our earth, both as to inherent heat and as to heat received from without. It seems to me that we may in this consideration find the real meaning of the comparatively limited extension of the Martian snows. It has been well pointed out by Professor Tyndall that for the formation of great glacial masses, not great cold only, but great heat also is required. The snows which fall on mountain slopes, to be compacted into ice and afterwards to form great glaciers, were raised into the air by the sun's heat. Every ice particle represents the action of that heat upon the particles of water at the surface of ocean, sea, or lake, or of wet soil. If the sun's heat suddenly died out, there would prevail an intense cold, and the snows and ice now existing would assuredly remain. The waters also of the earth would congeal. But no new snows would fall. The congealed seas viewed from some remote planet would appear unchanged. For they would not be covered with snow and broken ice, nor therefore white; but would consist of pure ice throughout, retaining the partial transparency and greenish colour of deep-sea water. No winds would disturb the surface of the frozen seas, for winds have their origin in heat, and with the death of the solar heat the winds would utterly die out also.

If we are to choose between these two explanations,— one that the snows and ice have not the great range we should expect, because the temperature is somehow raised despite Mars's greater distance to the same temperature which we experience, and the other that it is not heat but cold which diminishes the quantity of Martian snow, I conceive that there is every reason the case admits of for accepting the latter instead of the former explanation. As extreme cold would certainly prevent glacial masses from being very large and deep, simply because the stores whence the ice was gathered would be less, the snow caps of a very cold planet would vary as readily with varying seasons as those of a planet like our earth. For though less heat would be poured upon them with the returning summer, less heat would be required to melt away their outskirts.

I think we may fairly regard Mars as in all probability a somewhat old and decrepit planet. He is not absolutely dead, like our own moon, where we see neither seas nor clouds, neither snow nor ice, no effects, in fine, of either heat or cold. But I think he has passed far on the road towards planetary death,—that is, towards that stage of a planet's existence when at least the higher forms of life can no longer exist upon the planet's surface.

There is one peculiarity of the planet's appearance which seems strikingly to accord with this view that Mars holds a position intermediate between that of our earth and the moon,—as indeed we might fairly expect from his intermediate proportions. The seas of our earth cover nearly three-quarters of her entire globe. The moon has no visible water on her surface. If we examine the chart of Mars at page 167, we see that the seas and oceans of the planet are much smaller (relatively as well as actually) than are the seas of our own earth. I have carefully estimated their relative extent in the following simple but effective way. I drew a chart such as the above-mentioned, but on a projection of my own invention, in which equal surfaces on a globe are represented by equal surfaces on the planisphere. Then I cut out with a pair of scissors the parts representing land and the parts representing water (leaving the polar parts as doubtful), and carefully weighed these in a delicate balance. I found that they were almost exactly equal: whatever preponderance there was seemed to be in favour of the land. Thus, if we assume that, when in the same stage of planetary existence, Mars had as great a relative extent of water surface as our earth, or that about $\frac{72}{100}$ of the surface of Mars were originally water, we should have to admit that the water had so far been withdrawn into the planet's interior as to diminish the water-surface by $\frac{22}{100}$ (for there are now barely $\frac{50}{100}$). At a very fair assumption as to the slopes of the Martian sea-bottoms, it would follow that more than half the Martian water originally existing above the surface had been withdrawn into the interior, as the planet's mass gradually cooled.

I am aware the assumption above mentioned is in itself somewhat daring, and is not supported by direct evidence. But, since we have very strong reasons for considering that the moon once had seas, which have been withdrawn in the way suggested, and since Mars unquestionably holds a position midway between the earth and moon as to size and presumably as to age,* it seems not unreasonable to find in the character of her seas,—less extended relatively than the earth's, but, unlike the moon's, still existing,—the evidence that she has gone partially through the process through which the moon has long since passed completely.

I think it very likely that the recent discovery of two Martian satellites will lead many to look with more disfavour than ever on the idea that Mars may not at present be the abode of life. For moons seem so manifestly convenient additions to a planet's surroundings, as light-givers, time-measurers, and tide-rulers, that many will regard the mere fact that these conveniences exist as proof positive that they are at this present time subserving the purposes which they are capable of subserving. I would point out, however, that our own moon must have existed for ages before any living creatures, far less any reasoning beings, could profit by her light, or by the regularity of her motions, or by her action in swaying the waters of ocean. And doubtless she will continue to exist for ages after all life shall have passed away from the

* By age here I do not mean absolute age, but relative age. I speak of Mars and the Moon as older than the earth in the same sense that I should speak of a fly in autumn as older than a five-year-old raven.

earth. Again, there can be no question that our earth would present a most attractive scene if she were viewed from the moon, and would be a most useful ornament of the lunar skies. Yet we have every reason to believe that there is not a living creature on the moon at present to profit by her light. The case may well be the same (apart from the actual evidence that it *is* the same) with Mars. His satellites may long since have served most useful purposes to his inhabitants; but it by no means follows that because if there were inhabitants on Mars now the same purposes would still be subserved, therefore there are inhabitants there.

Let us, however, without considering the question whether the satellites of Mars serve such special purposes for creatures living on the planet, consider briefly the history of their discovery, their nature, and the laws of their motion around the planet.

Astronomers had long examined the neighbourhood of Mars with very powerful telescopes, in the hope of discovering Martian moons. But the hope had so thoroughly been abandoned for many years that the planet had come to be known as "moonless Mars." The construction, however, of the fine telescope which has been mounted at Washington, with an object-glass twenty-six inches in diameter, caused at least American astronomers to hope that after all a Martian moon or two might be discovered.

Taking advantage of the exceptionally favourable opportunity presented during the planet's close approach to our earth in the autumn of 1877, Prof. Asaph Hall, of the Washington Observatory, paid special attention to the search for Martian moons. At last, on August 16, 1877, he detected close by the planet a faint point of light, which he was unable to examine further at the time (to see if it behaved as a satellite, or as one of the fixed stars). But on the 18th he saw it again, and determined its nature. He also saw another still fainter point of light closer to the planet; and subsequent observations shewed that this object also was a satellite. During the next few weeks both the moons were observed as closely as possible, in fact, whenever weather permitted, and the result is that we now know the true nature of their paths.

In fig. 23 these paths are shown as they appeared in 1877. Of course the paths themselves are not seen; but if the satellites left behind them a fine train or wake of light, the shape of this train would be as shown in fig. 23. The satellites themselves could not be shown at all in a picture on so small a scale—the diameter of either would certainly be less than the cross-breadth of the fine elliptical line representing its track. The size of the planet is correctly indicated, and the true pose of the planet in 1877 is shown in the figure, his southern pole being somewhat bowed towards the earth. This is the uppermost pole; for the figure represents the planet and his satellites' orbits as they would appear in an astronomical telescope, which inverts objects.

The outer satellite is probably not more than ten miles or so in diameter, the inner one, perhaps, the same; but neither can be measured. In the most powerful telescopes they appear as mere points of light.

Fig. 23 —Mars and the paths of the Martian satellites as at present situated.

Nor is it easy to determine, from their lustre, or rather from their faintness, their true dimensions; for we cannot compare them directly in this respect with objects of known size, because their visibility is partly affected by the proximity of the planet, whose overpowering light dims their feeble rays. This remark applies with special force to the inner satellite.

The distance of the outer satellite from Mars's centre

is about 14,300 miles, from Mars's surface about 12,000 miles. The inner travels at a distance of about 5,750 miles from the centre, and about 3,450 miles from the surface of Mars.

The motions of the satellites as seen from Mars must be very different from those of our own moon. Thus, our moon moves so slowly among the stars that she requires nearly an hour to traverse a distance equal to her own apparent diameter. The outer moon of Mars traverses a similar distance—that is, not her own apparent diameter, but an arc on the stellar heavens equal to our moon's apparent diameter—in about two and a half minutes, while the inner moon moves so rapidly as to traverse the same distance in about forty seconds. To both moons, therefore, but to the inner in particular, Job's description of our moon as "walking in brightness" would seem singularly applicable, so far at least as the rapidity of their motions is concerned. Their brightness, however, cannot be comparable to our moon's. For notwithstanding their much greater proximity, it is easily shown that they must present much smaller discs, and being illuminated by a more distant sun, their discs cannot shine so brightly as our moon's. That is, not only are the discs smaller, but their intrinsic brightness is less. Assuming the outer moon to be ten miles, the inner fifteen miles in diameter, it is easily shown that the two together, if full at the same time, can hardly give one-twelfth as much light to Martians as our moon gives to us.

Yet there can be no doubt that the Martian moons must be (or *have been*) most useful additions to the Martian skies. They do not give a useful measure of time intermediate in length between the day and the year, as our moon does; for the outer travels round the planet in about thirty and a quarter hours, the inner in about seven and a half hours. Nor can they exert an influence upon the Martian seas corresponding to that exerted by our own moon in generating the lunar tidal wave. But their motions must serve usefully to indicate the progress of time, both by night and by day, for they must be visible by day unless very close to the sun. They must be even more useful than our moon in indicating the longitude of ships at sea, seeing that the accuracy with which a moon indicates longitude is directly proportional to her velocity of motion among the stars.

I have said that there does not seem to be any valid reason for considering that now is the accepted time with these moons; their services may have been of immense value in long past ages, and now be valueless for want of any creatures to be benefited by them. But those who not only believe that no object in nature was made without some special purpose, but that we are able to assign to each object its original purpose, should be well satisfied if they find reason for believing that, during millions of years long, long ago, the moons lately discovered by our astronomers were measuring time for past races of Martians, swaying tides in wider seas than those which now lave the shores of Martian continents, and enabling Martian travellers to guide their course over the trackless ocean and arid desert with far greater safety than can our voyagers by sea and land despite all the advances of modern science.

X.

THE PLANET JUPITER.

TWO or three years ago I had occasion to consider in the *Day of Rest* the giant planet Jupiter, the largest and most massive of all the bodies circling around the sun. I then presented a new theory respecting Jupiter's condition, to which I had been led in 1869, when I was visiting other worlds than ours. Since then, in fact within the last few months, observations have been made which place the new theory on a somewhat firm basis; and I propose now briefly to reconsider the subject in the light of these latest observations.

In the first place I would call the reader's attention to the way in which modern science has altered our ideas respecting time as well as space, though the change has only been noticed specially as it affects space. In former ages men regarded the region of space over which they in some sense had rule as very extensive indeed. This earth was the most important body in the universe, all others being not only made for the service of the earth, but being in all respects, in size, in range, and so forth, altogether subordinate to it. Step by step men passed from this to an entirely different conception of our earth's position in space. Shown first to be a globe within the domain of the heavenly bodies, then to be a globe subordinate to the sun, then to be a member of one family among thousands each with its ruling sun, then to belong to a galaxy of suns which is but one among myriads of millions of such galaxies, and lastly shown to the eye of reason, though not to direct observation, as belonging to a galaxy of galaxies itself but one among millions of the same order, which in turn belong to higher and higher orders endlessly, the earth has come to be regarded, despite its importance to ourselves, as but a point in space. The minutest particle by which a mathematician might attempt to picture the conception of a mathematical point, comparing that particle with any near object however large, a house, a mountain, the earth itself, would be but the grossest representation of a point, by comparison with the massive earth, when she is considered with reference to the universe of the fixed stars or rather to that portion of the universe, itself but a point in space, over which the survey of the astronomer extends.

All this has been admitted. Men have fully learned to recognise, though they are quite unable to conceive, the utter minuteness, one may say the evanescence, of their abode in space.

But along with the extension of our ideas respecting space, a corresponding extension has been made, or should have been made, in our conceptions respecting time. We have learned to recognise the time during which our earth has been and will be a fit abode for living creatures as exceedingly short compared with the time during which she was being fashioned into fitness for that purpose, and with the æons of æons to follow, after life has disappeared from her surface. This, however, is but one step towards the eternities to which modern science points. The earth is but one of many bodies of a system; and though it has been the custom to regard the birth of that system as if it had been effected, if one may so speak, in a single continuous effort (lasting millions of millions of years, mayhap, but bringing all the planets and their central sun simultaneously into fitness for their purpose), there is no reason whatever for supposing this to have been really the case, while there are many reasons for regarding it as utterly unlikely. It seems as though men could not divest themselves of the idea that our earth's history is the history of the solar system and of the universe. Precisely as children can hardly be brought to understand, for a long time, what history really means, how generation after generation of their own race has passed away, and how their own race has succeeded countless others, so science, still young, seems scarcely to appreciate the real meaning of its own discoveries. It follows directly from these that world after world like our earth, in this our own system or among the millions peopling space, has had its day, and that the systems themselves, on which such worlds attend, are but the existent representatives of their order, and succeed countless other systems which have long since served their purpose.

Yet, strangely enough, students of science continue for the most part to speak of other worlds, and other suns, and other systems, as though this present era, this "bank and shoal of time," were the sole period to which to refer in considering the condition of those worlds and suns and systems. It does not seem to occur to them that,—not possibly or probably, but most certainly,—myriads among the celestial bodies must be passing through stages preceding those which are compatible with the existence or support of life, while myriads of others must long since have passed that stage. And thus ideas appear strange and fanciful to them which, rightly apprehended, are alone in strict accordance with analogy. To consider Jupiter or Saturn as in the extreme youth of planetary existence, still glowing with such heat as pervaded the whole frame of our earth before she became a habitable world, still enveloped in cloud masses containing within them the very oceans of those future worlds, all this is regarded as fanciful and sensational. Yet those who so regard such theories do not hesitate to admit that every planet must once in its life pass through the fiery stage of planetary existence, nor are they prepared to show any reason why the stage must be regarded as past in the case of every planet or even of most of the planets. Seeing that, on the other hand, there are abundant reasons for believing that planets differ very widely as regards the duration of the various stages of their life, and that our earth is by no means one of the longest lived, we may very fairly expect to find among the planets some which are very much younger than our earth,—not younger, it will be understood, in years, but younger in the sense of being less advanced in development. When we further find that all the evidence accords with this view, we may regard it as the one to which true science points.

All that we know about the processes through which our earth has passed suggests the probability, I will even say the certainty, that planets so much larger than she is as are Jupiter and Saturn must require much longer periods for every one of those processes. A vast mass like Jupiter would not cool down from the temperature which our earth possessed when her surface was molten to that which she at present possesses in the same time as the earth, but in a period many times longer.

Supposing Bischoff to be right in assigning 340,000,000 years to that era of our earth's past, I have calculated that Jupiter would require about seven times and Saturn nearly five times as long, or about 2,380,000,000 and 1,500,000,000 years respectively, and by these respective periods would they be behind the earth as respects this stage of development. Suppose, however, on the other hand, that Bischoff has greatly overrated the length of that era—and I must confess that experiments on the cooling of small masses of rock, such as he dealt with, seem to afford very unsatisfactory evidence respecting the cooling of a great globe like our earth. Say that instead of 340,000,000 years we must assign but a tenth part of that time to the era in question. Even then we find for the corresponding era of Jupiter's existence about 238,000,000 years, and for that of Saturn's 150,000,000 years, or in one case more than 200,000,000 years longer, in the other more than 110,000,000 years longer than in our earth's case.

This relates to but one era only of our earth's past. That era was preceded by others which are usually considered to have lasted much longer. The earth, according to the nebular theory of Laplace, was once a mighty ring surrounding the sun, and had to contract into globe form, a process requiring many millions of years. When first formed into a globe she was vaporous, and had to contract—forming the moon in so doing—until she became a mass, first of liquid, then of plastic half solid matter, glowing with fire and covered with tracts of fluent heat. Here was another stage of her past existence, requiring probably many hundreds of millions of years. Jupiter and Saturn had to pass through similar stages of development, and required many times as many years for each of them. Is it then reasonable to suppose that they have arrived at the same stage of development as our earth, or indeed as each other.

Supposing for a moment that we were fully assured that Jupiter and Saturn had separate existence, hundreds of millions of years before our earth had been separated from the great glowing mass of vapour formerly constituting the solar system, and that having this enormous start, so to speak, they need not necessarily be regarded as now very greatly in arrear as respects development, or might even be in advance of the earth, it is altogether improbable that either of them, and far more improbable that both of them, are passing through precisely the same stage of development. If we knew only of two ships, that one had to travel from New York to London, and another from Canton to Liverpool, some time during the year, and that the one which had to make the longer journey was likely to start several weeks before the other, would it not be rather unsafe to conclude, when the former had entered the mouth of the Thames, that in all probability the other was sailing up the Mersey? Yet something like this, or in reality much wilder than this, is the reasoning which permits the student of science to believe, independently of the evidence, or altogether against all evidence, that Jupiter and Saturn are necessarily passing through the very stage of planetary existence through which the one planet we know much about is passing.

It seems to me that the student of science should be prepared to widen his conceptions of time even as he has been compelled to widen his conceptions of space. As he knows that the planets are not, as was once supposed, mere attendants upon our earth and belonging to her special domain in space, so should he understand that neither do the other planets appertain of necessity to the domain of time in which our earth's existence has been cast, or only do so in the same sense that like her they occupy a certain domain in space, not her domain, but the sun's. Their history in time, like hers, doubtless belongs to the history of the solar system, but the duration of that system enormously surpasses the duration of the earth as a planet, and immeasurably surpasses the duration of that particular stage of life through which she is now passing.

Prepared thus to view the other planets independently of preconceived ideas as to their resemblance to our own earth, we shall not find much occasion to hesitate, I think, in accepting the conclusion that Jupiter is a very much younger planet.

We have seen already that the enormous mass of Jupiter, surpassing that of our earth 340 times, is suggestive of the enormous duration of every stage of his existence, and therefore of his present extreme youth. His bulk yet more enormously exceeds that of our earth, as, according to the best measurements, no less than 1230 globes, as large as our earth, might be formed out of the mighty volume of the prince of planets. In this superiority of bulk, nearly four times greater than his vast superiority of mass, we find the first direct evidence from observation in favour of the theory that Jupiter is still intensely hot. How can a mass so vast, possessing an attractive power in its own substance so great that, under similar conditions, it should be compressed to a far greater degree than our earth, and be, therefore, considerably more dense, come to be considerably rarer? We no longer believe that there is any great diversity of material throughout the solar system. We cannot suppose, as Whewell once invited us to do, that Jupiter consists wholly or almost wholly of water. Nor can we imagine that any material much lighter than ordinary rocks constitutes the chief portion of his bulk. We are, to all intents and purposes, forced to believe that the contractive effect due to his mighty attractive energy is counteracted by some other force. Nor can we hesitate, since this is admitted, to look for the resisting force in the expansive effects due to heat. We know that in the case of the sun, where a much mightier contractive power is at work, a much more intense heat so resists it that the sun has a mean density no greater than Jupiter's. We have every reason, then, which bulk and mass can supply, to believe that Jupiter is far hotter than the earth—that in fact, as the sun, exceeding Jupiter more than 1000 times in volume, is many times hotter than he is, so Jupiter, exceeding our earth 1200 times in volume, is very much hotter than the earth.

But when we consider the aspect of Jupiter we find that similar reasoning applies to his atmosphere. The telescope shows Jupiter as an orb continually varying in aspect, so as to assure us that we do not see his real surface. The variable envelope we do see presents, further, all the appearance of being laden with enormously deep clouds. The figure (24) shows the planet as seen by Herr Lohse on February 5, 1872, and serves to illustrate the rounded clouds often seen in Jupiter's equatorial zone, as though floating in the deep atmosphere there. Although rounded clouds such as these are not constantly present, they are very often seen; their appearance, even on a few occasions only, would suffice for the argument I now propose to draw from

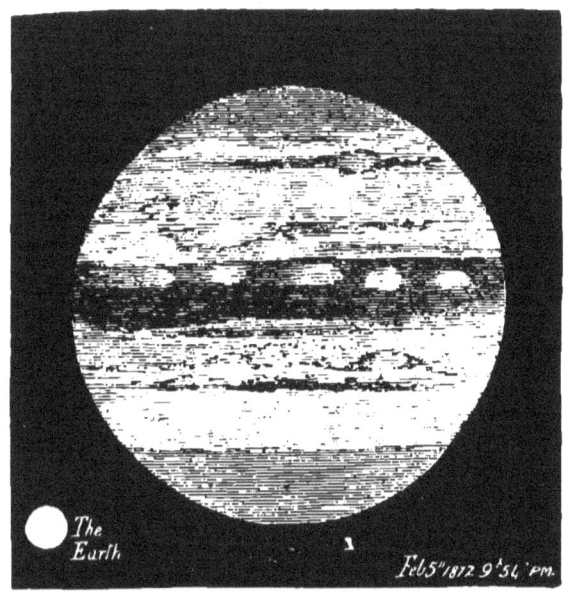

Fig. 24.—The Planet Jupiter.

them. It is impossible to regard them as *flat* round clouds. Manifestly they are globular. Now they may not be quite as deep as they are long, or even broad, but supposing them only half as deep as they are broad, that would correspond to much more than a third of the diameter of our earth, shown in the same picture. The atmosphere in which they float would necessarily be deeper still, but that depth alone would be about 3,000 miles. Now an atmosphere 3,000 miles deep under the tremendous attraction of Jupiter's mass would be compressed near its base to a density many times exceeding that of the densest solids *if* (which of course is impossible) it could remain in the gaseous form with such density. The fact, then, that an atmosphere, certainly gaseous, exists around Jupiter to this enormous depth at least, proves to demonstration that there must be some power resisting its attractive energy; and again, we have little choice but to admit that that power is no other than the planet's intense heat.

As we extend our scrutiny into the evidence from direct observation, we find still other proofs independent of those just considered. One proof alone, be it remembered, is all that is required, but it will be found that there are many.

We have found reasons for believing that the planet Jupiter is expanded by heat in such sort that the contractive or condensing power of his own mighty attractive energy is overcome. We know certainly that, regarding the planet we see as a whole, its globe is of very small density. We have every reason to believe that it is made of the same materials, speaking generally, as our earth. We know that its mass as a whole possesses many times the gravitating power of our earth's mass. It is highly probable, therefore, that the condition of its substance is very different from that of our earth's substance. And as we know of no cause save heat which could keep the planet in this state, it is altogether probable that the planet is extremely hot. The argument, be it noticed, is independent of that based on the probability that Jupiter, owing to his enormous mass, has not cooled nearly so much as our earth has.

We then noticed another very powerful argument, similar in kind, but also quite independent, derived from the aspect of the planet. Jupiter's appearance indicates clearly that he has a deep cloud-laden atmosphere, and we know that such an atmosphere, if of the same temperature as our earth's, would be compressed enormously, whereas the observed mobility of Jupiter's cloud-envelope, and other circumstances, indicate that this enormous compression does not exist. We infer, then, that some cause is at work expanding the atmosphere; and we know of no other cause but heat which could do so effectively.

But now let us consider certain details which the telescope has brought to our knowledge.

In the first place, a number of circumstances indicate a tremendous activity in that deep cloud-laden air.

The cloud-belts sometimes change remarkably in appearance and shape in a very short time. Mr. Webb, in his excellent little treatise, "Celestial Objects for Common Telescopes," gives instances from the observations of South, which I here translate into non-technical terms:—On June 3, 1839, at about nine in the evening, South saw with his large telescope, just below the principal belt of Jupiter, a spot of enormous size. It was dark, and therefore probably represented an opening in a great cloud-layer by which a lower or inner layer was brought into view. (For though the planet's real globe may be so intensely hot as to emit a great deal of light, it is probable that most of the light so emitted is concealed by the enwrapping cloud masses, and that the greater part of the light we receive from the planet is reflected sunlight; so that the inner cloud-layers would be the darker.) South estimated this spot as about 20,000 miles in diameter. "I showed it," he says, "to some gentlemen who were present; its enormous extent was such that on my wishing to have a portrait of it, one of the gentlemen, who was a good draughtsman, kindly undertook to draw me one; whilst I, on the other hand, extremely desirous that its actual magnitude should not rest on estimation, proposed, on account of the scandalous unsteadiness of the large instrument, to measure it with my five-feet telescope."

"Having obtained for my companion the necessary drawing instruments, I went to work, he preparing himself to commence his." But on looking into the telescope, South was astonished to find that the large dark spot, except at its eastern and western edges, had become much whiter than any of the other parts of the planet; and thirty-four minutes after these observations had commenced, "these" [query three?] "miserable scraps were the only remains of a spot which but a few minutes before had extended over at least 20,000 miles, —or two and a half times the diameter of the earth."

The cloud envelope, then, of Jupiter is certainly not in a state of quiescence. Of course we need not suppose that winds had carried cloud masses athwart the tremendous opening seen by South. That would imply a motion of 10,000 miles in the half-hour or so of observation,—supposing contrary winds to have rushed towards the centre,—or double that velocity if the entire breadth of the spot had been traversed in that time. A velocity of 20,000 miles, and still more of 40,000 miles per hour, may fairly be regarded as incredible. It would exceed more than a hundred-fold (taking the least number) the velocity of our most tremendous hurricanes. And although the solar hurricanes would seem to have a velocity, at times, of 300,000 or 400,000 miles per hour, we have no reason for supposing that winds of tens of thousands of miles per hour could be raised in the atmosphere of Jupiter. As I have said, however, it is not necessary to suppose this. We may conceive that clouds had formed very rapidly at the higher elevation where before they had been wanting. Clouds may form as readily and quickly over an area a thousand miles across as over an area two or three miles across. Indeed Webb, referring to such changes as South witnessed, says that Sir J. Herschel once observed a cloud-bank in our own air, which formed so rapidly that it crossed the sky at the rate of 300 miles an hour, not moving, of course, at that rate, but being formed along different parts of its apparent progress almost simultaneously, so as to appear to travel with this enormous velocity.

But now I wish the reader specially to notice how this observation of South's may serve to explain another, equally remarkable and at first sight far more perplexing; and how, when the two observations are brought together, a very singular piece of information is obtained respecting Jupiter's cloud-envelope.

Let *a b*, fig. 25, represent the great dark space seen by South, just below the principal belt, and let us suppose the planet turned round until this dark space, or rather this opening in the planet's outer cloud-envelope, is brought to the edge as at *a c d*, fig. 26. Then this opening would really cause a depression in the planet's outline at *d c*, the shaded part being depressed. The depression might not be observable in any ordinary telescope. For at the edge of Jupiter the features of

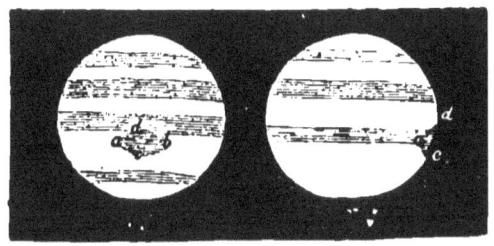

Fig. 25.   Fig. 26.

the belt are generally lost, and the outline is at all times smoothed in appearance by that peculiarity of vision which makes all bright objects seen on a dark background appear somewhat larger than they actually are. (This is *really* due to a fringing, as it were, of the image on the retina of the eye.) But though the depression might not be recognisable, it would exist, and, as we shall presently see, it might be detected in another way than by being actually seen. When the clouds formed which concealed the spot,—we do not know how quickly, but certainly in less than thirty-four minutes,—the depression, had the spot been at the edge, would have been removed. This change, however, like the existence of the depression, would doubtless not have been discernible by ordinary vision.

Now, let us consider the second observation mentioned above.

On Thursday, June 26, 1828, the second satellite of Jupiter was about to make a passage across the planet's face. It was observed, just before this passage or transit began, in the position shown in fig. 27 by the late Admiral Smyth. He was using an excellent telescope. It gradually made its entry, looking for a few minutes like a small white mountain on the edge of the planet, and finally

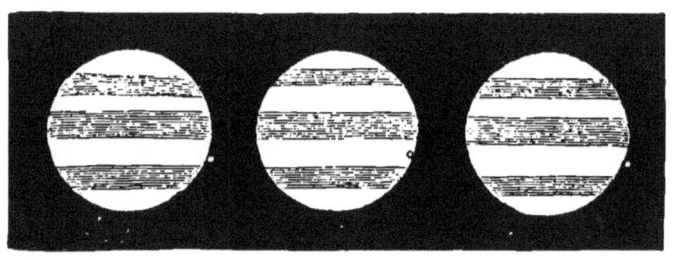

Fig. 27.   Fig. 28.   Fig. 29.

disappeared. The reader must understand that the moon was not hiding itself behind the planet, but was on *this* side of it, and simply lost to view because its brightness was the same, or very nearly the same, as that of the planet's edge. (Its place is shown in fig. 28, but of course the little dark ring was not so seen.) "At least 12 or 13 minutes must have elapsed," says Smyth, "when, accidentally turning to Jupiter again, to my astonishment I perceived the same satellite outside the disc," as shown in fig. 29. It remained visible there for at least four minutes, and then suddenly vanished. To show that the observation was not due to any local or personal peculiarity, it is only necessary to mention that two other astronomers, Mr. Maclear at Biggleswade, and Dr. Pearson at South Kilworth, observed the same extraordinary behaviour of Jupiter's second satellite. The three telescopes are thus described by Admiral Smyth,—

> Mr. Maclear's, 3¼ inches in opening, 3½ feet long;
> Dr. Pearson's 6¾ inches in opening, 12 feet long;
> Adm. Smyth's 3⅓ inches in opening, 5 feet long;

all good observing telescopes. Now, of course, the satellite did not really stop. Nothing short of a miracle could have stopped the satellite, or, if the satellite could have stopped, have set it travelling on again as usual. For the satellite did not lose one mile, or change its velocity by the thousandth part of a mile per hour or even per annum.

But suppose such a change had taken place at the edge of Jupiter as we have seen would certainly have taken place there if the changes affecting the spot which South saw had occurred to a region at the edge, as in fig. 26, instead of the middle, as in fig. 25. Then Smyth's observation would be perfectly explained. We require, indeed, to suppose the change occurring in a different order, the outer cloud-layer being in the first instance well-developed and very rapidly becoming dissipated, so that the outline which had at first been at its usual level, was very rapidly depressed to the inner cloud-layer. But, of course, if the rapid formation of clouds by condensation can occur on Jupiter, so also can the rapid dissipation occur, especially at that particular part where Smyth saw the satellite behave so strangely. For that part is being carried, by the planet's swift rotation, into sunlight, and the extra heat to which it is thus exposed might readily effect the dissipation of widely extended cloud strata, supposing the temperature near that critical value at which clouds form or are dissipated.

Here, then, is an explanation of a phenomenon which otherwise seems utterly inexplicable. The explanation requires only that a process like one which has been observed to occur on Jupiter's disc should occur at a part of his surface forming at the moment a portion of his outline. If we had never known of such changes as South and other observers have noted in the markings of Jupiter, we should be compelled by Smyth's observation to admit their possibility. If we had never known of Smyth's observation we should be compelled by South's to admit that such a change of outline as is indicated by Smyth's observation must be possible,— must, in fact, occur whenever cloud-masses form or are dissipated over wide areas at the apparent edge of the planet. When we have both forms of evidence it seems altogether unreasonable to entertain any further doubt on this point.

But Smyth's observation, thus interpreted, indicates an enormous distance between the outer and inner cloud-layers which formed the planet's edge near the satellite in figs. 28 and 29 respectively. I find after making every possible allowance for errors in his estimate of time, not taken it would seem from his observatory clock, that the distance separating these cloud-layers cannot have been less than 3,500 miles, or not far from half the diameter of our earth. It is the startling nature of this result which perhaps deters many from accepting the explanation of Smyth's observation here advanced. But there is no other explanation. The satellite cannot have stopped in its course; Jupiter cannot have shifted his place bodily; the satellite was on this side of the planet, —therefore no effects of the planet's atmosphere on the line of sight from the planet can help us; three observers at different stations saw the phenomenon,—therefore neither effects of our earth's atmosphere nor personal peculiarities can account for the strange phenomenon. "Explanation is set at defiance," says Webb; "demonstrably neither in the atmosphere of the earth nor Jupiter, where and what could have been the cause?" The explanation I have advanced is the only possible answer to this question.

I might occupy twenty times the space here available to me in detailing various other phenomena all pointing in the same way,—that is, all tending to show that Jupiter is a planet glowing with intense heat, surrounded by a deep cloud-laden atmosphere, intensely hot in its lower portions, but not necessarily so in the parts we see, and undergoing changes (consequences of heat) of a stupendous nature, such as the small heat of the remote sun, which shines on Jupiter with less than the 27th part of the heat we receive, could not by any possibility produce. But partly because space will not permit, partly because most of these phenomena have been described in my "Orbs Around Us," and "Other Worlds," I content myself by describing a singular observation recently made, which, with South's and Smyth's, seems to place the theory I have advanced beyond the possibility of doubt or cavil.

Mr. Todd of Adelaide has recently obtained for his observatory a fine 8-inch telescope by Mr. Cooke. With this instrument, mounted in December, 1874, he has made many valuable observations of the motions of Jupiter's satellites. Ordinarily, of course, the entry of each satellite on the planet's face and the egress therefrom, the disappearance of each satellite behind the planet or in the planet's shadow (not necessarily the same thing) and the reappearance, are effected in what may be called the normal way; and Mr. Todd's experience in this respect has been like that of other observers. But on two occasions he and his assistant, Mr. Ringwood, observed that a satellite, when passing behind the planet's edge, did not disappear at once, but remained visible as if seen through the edge, for about two minutes. The same satellite behaved thus on each occasion,—viz. the satellite nearest the planet. As this satellite travels at the rate of about 645 miles per minute, it would follow that the satellite was seen through a depth of nearly 1300 miles, or, after making all possible allowance for optical illusions, some 900 or 1000 miles. The effect of refraction cannot then be great in the air of Jupiter, to this depth below the usual limit of the upper clouds,— for otherwise the satellite would have been altogether distorted. And this very fact, that for 1000 miles or so below the highest clouds the change of atmospheric density is not sufficient to produce any noticeable refractive effects, implies that the true base of the atmosphere of Jupiter lies very far lower yet—perhaps many hundreds of miles lower.

If the reader now look again at the picture at page 201, he will understand, I think, how those great round white clouds in the chief belt,—clouds thousands of miles long and broad,—are probably hundreds of miles deep also, and float in an atmosphere still deeper.

All that we know about Jupiter, in fine, from direct observation, as well as all that we can infer respecting the past history of the solar system, tends to show that he is still an extremely young planet. He is the giant of the solar family in bulk, and probably he is far older than our earth in years; but in development he is, in all probability, the youngest of the sun's family of planets, and certainly far younger than the earth on which we live.

XI

THE RINGED PLANET SATURN.

VERY different from the ruddy planet which approached so closely to him in November, 1877, is Saturn, the ringed world, the most wonderful of all the planets if the complexity of the system attending on him is considered, and in size inferior only to the giant Jupiter.

It will have been noticed, perhaps, by those who are familiar with the aspect of the planets, that the contrast between Mars and Saturn during their late approach to us was not only greater than usual, but greater than was to be expected even when account was taken of the unusual lustre of Mars. I have often wondered whether the ancient astronomers where ever perplexed by the varying lustre of Saturn. They recognised the fact that Mars has an orbit of great eccentricity (see the picture of the orbits of Mars, Venus, etc., at page 156);

and there was nothing in the varying lustre of Mars which could not be perfectly well explained by his known variations of distance, whether the Ptolemaic or the Copernican system were accepted. But with Saturn the case is different. His distance at successive returns to our midnight skies is subject to moderate changes only. Yet his brilliancy varies in a remarkable manner. We now know that those changes are due to the opening and closing of that marvellous system of rings which renders this planet the most beautiful of all the objects of telescopic observation which the heavens present to us. When the edge of the rings is turned towards the earth, we see only the most delicate thread of light on either side of the planet's disc. But when the rings are opened out to their full extent they reflect towards us as much light as we receive from the disc. At such times the planet presents a much more brilliant appearance than when the ring is turned nearly edgewise; in fact to the naked eye he seems very nearly twice as bright. Now at present the rings are turned nearly edgewise towards the earth. In July and August, 1869, the planet presented in the telescope the appearance presented in fig. 30, where it will be seen that the shorter axis of the oval into which any one of the ring-outlines is thrown is nearly equal to half the larger axis. Since then the rings have been slowly closing up; and at present the rings are so little open that the corresponding shorter axis, if it could be directly seen, would appear to be about one-sixteenth only of the larger axis. The rings were turned exactly edgewise towards the sun at two in the afternoon, on St. Valentine's day, 1878, according to calculations which I made in 1864, and

Fig. 30.—The planet Saturn in July and August, 1869.

published in a table under the head "Passages of the Rings plane through the Sun between the years 1600 and 2000," in my treatise entitled "Saturn and its System." The *Nautical Almanac* for 1878, indeed makes the passage of the rings plane through the sun occur somewhat earlier, stating that at noon on February 14 the sun's centre would pass south of the ring's horizon by about one-fifth of its apparent diameter (as seen by us). But my own calculation took into account certain small details which, in matters of this sort, the *Nautical Almanac* computers neglect. After all, it mattered very little to terrestrial observers whether the sun's light passed from the northern to the southern side of the rings a few hours earlier or later: the moment when it passed could not possibly be observed, even if it had occurred during the night hours. In the present instance it occurred at midday, and unfortunately none of the interesting phenomena presented in powerful telescopes when the rings are turned edgewise to the sun or earth could be observed, for they occurred when Saturn and the sun were nearly in the same part of the heavens, and when the planet therefore was utterly lost in the splendour of the solar rays.

But now let us briefly consider what is known or may be surmised respecting the noble planet which was so far outshone in November, 1877, by the comparatively minute orb of Mars.

Saturn travels at a distance from the sun exceeding rather more than nine and a half times that of our own earth. The second figure of orbits (see page 157) shows the wide span of his orbit compared with the earth's, and yet it will be seen that the orbits of Uranus and Neptune, planets unknown to the ancients, are so wide that the path of Saturn becomes in turn small by comparison.

Saturn has a globe about 70,000 miles in diameter, where it bulges out at the equator; but he is somewhat flattened at his poles, so that his polar diameter is about 7000 miles less than the equatorial diameter. In volume he exceeds our earth about 700 times; but in mass only about ninety times: for his mean density is but about $\frac{13}{100}$ of the earth's. In fact, if we could imagine an ocean of water wide enough and deep enough for the planets to be all set in it, Saturn would float with about one-fourth of his bulk out of water,—always supposing that no change took place in his density directly after he was immersed. Saturn, indeed, would float highest of all the planets, or rather all of them would sink except Saturn and Neptune, and Saturn would float higher than Neptune. Uranus would just sink. Jupiter is half as heavy again as he should be to float. All the terrestrial planets, Mercury, Venus, the Earth, and Mars, would go to the bottom at once.

It is almost impossible to regard any feature of Saturn as better deserving to be considered first than his ring system. Yet for the sake of preserving a due sequence of ideas we must first consider his globe.

We find ourselves at once in presence of difficulties like those we encountered when we considered the planet

Jupiter. How is it that the mighty mass of a planet like Saturn, constructed, we have every reason to believe, of materials resembling those which constitute our earth, has so failed to gather in its substance that the mean density is much less than that of the earth's globe? It must be remembered that gravity prevails throughout the frame of Saturn as throughout our earth's frame. Every particle of that enormous globe is drawn towards the centre with a force almost exactly the same as would be exerted by the attraction of the entire mass of that portion of the planet which lies nearer to the centre than the particle, if this mass were collected at the centre. But this is not all. It is not merely the attraction exerted on each particle of Saturn's mass which has to be considered, but the entire *weight* of all the superincumbent matter. The distinction between attraction and weight, by the way, is very commonly overlooked in considering the planets' interiors. I think it was Sir David Brewster who argued that as attraction can easily be shown to diminish downwards towards the centre, it is possible to conceive that the interior of a planet may be hollow. The error is readily perceived, if we take a familiar instance where the attraction is the same yet the effect of pressure very much greater. (Without voyaging to the centre of the earth, which is troublesome, and certainly not a familiar experience, we cannot reach places where the attraction of gravity is greatly less than at the surface.) Take a massive arch of brickwork: the bricks near the top are subject to the same attraction as those belonging to the foundation; but the pressure to which the foundation bricks are exposed is very much greater than that affecting the upper bricks. So again with a deep sea: the particles at the bottom of such a sea are subject to no greater attraction than the particles near the top; but we know that a strong hollow case of metal which near the top of such a sea would be scarcely pressed at all, and would suffer no change of shape, will be crushed perfectly flat under the tremendous pressure to which it will be exposed when sunk to the bottom.

There is, in fact, no escape from the conclusion that the interior portions of a planet like Saturn or Jupiter, nay, even of a body like our earth or the moon, must be subject to tremendous pressure, a pressure exceeding many hundred-fold the greatest which we can obtain experimentally, and that under that enormous pressure the density of the materials composing those central parts must be increased. How is it then that Saturn is of much smaller density than the earth? I can imagine no other explanation at once so natural and so complete as this, that an intense heat pervading the entire frame of the planet enables it to resist the tremendous pressure due to mere weight. The planet's mass is expanded by the heat; large portions which at ordinary temperatures would be solid are liquified or even vaporised; matters which are liquid on our earth are vaporised; and, in fine, the planet assumes (as seen from our distant station) the appearance of being very much larger than it really is. We measure not the true globe, which, for aught that is known, may be exceedingly dense, but the dimensions of cloud-layers floating high in the planet's atmosphere.

In describing Jupiter, I considered the changes which have been noticed in that planet's outline, and observed that it is impossible adequately to explain the evidence, without assuming that the changes of outline are real. The outline is not that of a solid globe, however, but of cloud-layers surrounding such a globe, and probably at a great distance from its surface.

In Saturn's case we have very singular evidence to the same purpose. It was observed by Sir W. Herschel in April, 1805, that Saturn occasionally appears distorted, as though bulging out in the latitudes midway between the pole and the equator of the planet. Fig. 31 represents the appearance of the planet so far as shape is concerned, but the ring was not, when Sir W. Herschel observed it, so narrow as it is shown in fig. 31. In fact the ring had been turned edgewise to the earth two years before; and when Herschel noticed the abnormal appearance of

Saturn, the rings had begun to open out, though their outermost outline was still far within the regions of the planet which seemed to project as shown. Fig. 31 in fact represents Saturn as seen by Schröter in 1803, when, as he said, the planet did not seem to present a truly

Fig. 31.—Saturn's square-shouldered aspect.

spheroidal figure. Herschel tested his observations by using several telescopes of different dimensions,—ten, seven, twenty, and forty feet in length. In 1818, when the rings were scarcely visible, Kitchener saw the planet as shown in fig. 31, or "square-shouldered," as some have called it. On one occasion the present Astronomer-

Royal saw the planet of that figure. In January, 1855, Coolidge, using the fine refractor of the Cambridge U.S. Observatory, noticed that the equatorial diameter was not the greatest; on the 9th the planet seemed of its usual shape; but on December 6, Coolidge writes, "I cannot persuade myself that it is an optical illusion which makes the maximum diameter of the ball intersect the limb half-way between the northern edge of the equatorial belt, and the inner ellipse of the inner bright ring." This was at a time when the rings were nearly at their greatest opening; so that, including Schröter's observation, we have Saturn out of shape when his ring has presented every shape between that shown in fig. 30 and that shown in fig. 31. Again, in the report of the Greenwich Observatory for 1860-61, when the ring was nearly closed, it is stated that "Saturn has *sometimes* appeared to assume the square-shouldered aspect." Lastly, the eminent observers, G. Bond and G. P. Bond, father and son, have seen Saturn abnormally shaped, flattened unduly in the north polar regions in 1848, when the ring was turned edgewise towards us, and unsymmetrical in varying ways in 1855-57, when the ring was most widely opened.

Yet the planet's outline is usually a perfect oval, and has been shown to be so by careful measurements effected in some instances by the same observers, who, making equally careful measurements, have found the planet to be distorted.

Does it not seem abundantly clear that the great cloud-layers which float in the atmosphere of Saturn have a widely varying range in height, and that therefore as we see and measure the outline of the cloud-layers, we see and measure in effect a planet which is variable in figure? This seems so natural and complete an explanation of the observed peculiarities that it appears idle on the one hand to reject the evidence of some among the most skilful observers who have ever lived; or, on the other, to imagine that the solid frame of the planet has undergone changes so tremendous as would be involved by the observed variations of outline if they really signified that a solid planet had changed in shape.

The mighty globe of Saturn turns upon its axis nearly as quickly as Jupiter. It will be remembered that the Jovian day lasts only $9\frac{1}{2}$ of our hours, and as the diameter of Jupiter is about ten times the earth's, the equatorial parts of the giant planet travel some twenty-six or twenty-seven times as fast as those of our own earth, which move (rotationally) at the rate of more than a thousand miles an hour. Saturn's equatorial parts do not move quite so fast,—in fact, in this respect, Jupiter comes first of all the members of the solar system, including the sun himself. Saturn's equatorial circuit being almost nine times the earth's, while his day is little more than five-twelfths of the earth's, it follows that his equatorial parts move twelve-fifths of nine times, or nearly twenty-two times faster than the earth's. Their actual rotational rate is rather more than 22,000 miles an hour, or 367 miles a minute, or more than six miles a second. This is a wonderful rate of motion. It always seems to me one of the most striking results of modern astronomical research that we have to recognise in bodies like that dull looking star,—the heavy slow-moving Saturn, as the ancients called him,*—motions of such tremendous swiftness. The planet is not only rushing bodily along through space with a velocity of nearly six miles per second, but his equatorial parts are being carried round with a velocity somewhat exceeding six miles per second. (The coincidence must be regarded as accidental, but it has this curious effect, that the equatorial parts of Saturn near the middle of the disc we see are actually almost at rest with respect to the sun, being carried forward with the planet at the rate of about five miles and nine-tenths per second, and backward round the planet at the rate of

* Who assigned to him, as his representative metal, lead—a metal "heavy, dull, and slow," as Don Armado puts it, in "Love's Labour Lost."

about six miles and one-tenth per second. In fact there are always two points on the disc which are almost exactly at rest with respect to the sun, viz., those two points north and south of the equator where the rotational velocity is about five miles and nine-tenths per second, the velocity of Saturn in his orbit.)*

But let us turn from the contemplation of Saturn's globe, interesting though it undoubtedly is, to study those marvellous objects, the Saturnian rings.

The history of their discovery is interesting, but must not here detain us long. Briefly, it runs as follows :—

Galileo, in July, 1610, observing the planet Saturn with a telescope not powerful enough to show the rings, imagined at first that Saturn had two companion planets, one on either side of him, as though helping the planet along upon his road. (From a table relating to the rings, in my treatise on "Saturn and its System," the aspect of the ring, at the time of any such observation, can at once be inferred. In the present case, for example, it will be seen from the table that the rings were closing up as the

* Attention has lately been called, by the astronomers of the Washington Observatory, to the fact that the statement usually made in our books of astronomy, that Sir W. Herschel's latest determination of Saturn's rotation period was 10h. 29m., is incorrect. His only determination of the period gave 10h. 16m. 44s. for the Saturnian day.

time of their disappearance, December 28, 1612, drew near.) A year and a half later, Galileo looked again at Saturn, and lo! the companion planets were gone. He was perplexed beyond measure. "What is to be said concerning so strange a metamorphosis?" he asked. "Are the two lesser stars consumed after the manner of the solar spots? Have they vanished or suddenly fled? Has Saturn, perhaps, devoured his own children? Or were the appearances indeed an illusion or fraud, with which the glasses have so long deceived me, as well as many others to whom I have shown them? Now, perhaps, is the time come to revive the well-nigh withered hopes of those who, guided by more profound contemplations, have discovered the fallacy of the new observations, and demonstrated the utter impossibility of their existence. I do not know what to say in a case so surprising, so unlooked for, and so novel. The shortness of the time, the unexpected nature of the event, the weakness of my understanding, and the fear of being mistaken, have greatly confounded me."

Hevelius was similarly perplexed by the constantly vayring appearance of the planet. "Saturn," he informed his contemporaries, "presents five various figures to the observer, to wit—first, the mono-spherical; secondly, the tri-spherical; thirdly, the spherico-ansated; fourthly, the elliptico-ansated; fifthly, and finally, the sphericocuspidated;" of which we can only say, like Mr. Gilbert's Ferdinando, that "we know it's very clever; but we do not understand it."

It was not till 1659 that Huyghens, using a telescope *forty yards long*, was able to make out the real meaning of the appendages which had so perplexed Galileo and Hevelius. He announced to the world, in an anagram, his discovery that Saturn is girdled about by a flat ring nowhere touching the planet.

Huyghens also discovered the largest of Saturn's moons. He looked for no more, having the idea that, since six planets and six moons were now known, no more moons existed.

In 1663 the Brothers Ball discovered that the rings are divided into two, or, at any rate, that a broad black stripe, such as is shown in fig. 30, separates the outer portion of the ring from the inner. Two years later these observers saw the stripe on the northern side of the rings, when the rings had so shifted in position that observers saw their southern side. Dominic Cassini recognised a corresponding stripe on the southern side. This was regarded as proving that there is a real division between the rings. The width of the gap thus separating the outside of the inner ring from the inside of the outer cannot be less than 1,600 miles.

Cassini also detected another Saturnian moon in

October, 1671, and, later, he discovered three others, making five Saturnian moons in all.

Sir W. Herschel observed the rings with great care. He confirmed the discovery of the great division between the rings; but rejected the idea which was beginning to be entertained in his time, that there are many divisions. He found reasons for suspecting, but never actually proved, that the outer ring turns round in about $10\frac{1}{2}$ hours.

He also detected two small moons close to the outer ring. One other moon, detected independently by Bond at the Harvard Observatory, Cambridge, U.S., and by Lassell in this country in 1848, completes the set of eight moons now known to revolve around the planet Saturn. We need not here say much more about these moons, saving, perhaps, to note that the span of the entire Saturnian system of moons amounts to about 4,400,000 miles, nearly double that of the Jovian system. This is the largest system of satellites known to us. It is wonderful to reflect, when we look at the dull, slow-moving Saturn, that not only is the planet itself 700 times larger than the earth, not only is it girdled about by a ring system having a span exceeding more than 20 times the diameter of this earth on which we live, but that the entire span of the system over which that distant planet rules exceeds more than eighteen-fold the distance separating our earth from the moon.

Return we now, however, to the consideration of the Saturnian ring-system.

In 1850 a singular discovery was made. It was found by Bond, in America, and, a few days later, independently, by Dawes, in England, that inside the inner bright ring there is a dark ring almost as wide as the outer bright ring. One of the strangest circumstances about this inner ring is that where it crosses Saturn's disc the outline of the planet can be distinctly traced *through* the dark ring, which is thus, in a sense, a semi-transparent body. I say "in a sense," because it does not follow that it really consists of semi-transparent matter any more than it follows from our being able to see through a gauze veil that the individual threads forming the gauze are made of a semi-transparent material.

On examining recorded observations of the planet evidence was found that this dark ring is not, as was at first supposed, a recent formation. Where it crosses Saturn it had been mistaken in former times for a dark belt.

It had always been supposed that the rings are solid, or at any rate continuous bodies. The younger Cassini, indeed, ventured to express doubts on the subject, but with this solitary exception, no suspicion had ever existed among astronomers that the rings are otherwise than continuous, until the discovery of the dark ring.

When the singular fact was discovered that the body of the planet can be seen through the slate-coloured ring, the solidity of *this* ring, at any rate, began naturally to be questioned. The idea was suggested that this formation may be fluid. Mathematicians applied rigorous processes of investigation to the question whether a fluid ring can possibly exist in such a position. The inquiry led to a re-examination of the whole subject of the ring-system and its stability. Mathematicians took up the question where Laplace had left it more than half a century before. He had decided that solid rings might, under certain conditions, revolve around a planet without being broken. But his inquiry had not been carried to a conclusion. Now, when the work was completed, it was found that the requisite conditions are certainly not fulfilled by the Saturnian ring-system. The rings should be situated eccentrically, and heavier at one side than the opposite. In fact they should have a perceptible "bias." They exhibit, on the contrary, the most perfect symmetry of figure—this symmetry, indeed, constitutes the great charm of Saturn's telescopic appearance; and although, occasionally, the ball has not seemed to be quite in the middle of the ring-system, the displacement has never approached that which theory requires.

The conclusion to which mathematicians arrived was accordingly the following:—

The rings may be held to be formed of a multitude of tiny satellites, travelling nearly in one plane, each pursuing its own course around Saturn, according to the laws of satellite motion, though of course disturbed by the attraction of its fellow-satellites.

We owe this theory principally to the labours of Professor J. Clerk Maxwell, who gained the Adams Prize offered by the University of Cambridge for the best mathematical essay upon the conditions under which a ring-system such as Saturn's can exist. But Professor Pierce, of America, had (somewhat earlier) supplied a complete refutation of the idea that the rings are solid and continuous bodies.

When the rings are fully open, as in fig. 30, the Saturnian system affords as charming an object for telescopic observation as the astronomer can desire. The rings are then exhibited in their full beauty. The divisions, the dark ring, and the strange shading of the middle ring, can be well seen in a telescope of adequate power. The telescopic view is still more interesting when (as in fig. 30) the planet throws a well-marked shadow upon the rings.

But perhaps the most beautiful of all the features which Saturn presents to the telescopist is the strange variety of colour to be observed upon his surface, and upon that of the rings. Mr. Browning, the eminent optician, thus describes the colours which the planet presents in his 12-inch reflector:—

"The colours I have used," he says, referring to a painting of the planet, "were—for the rings, yellow-ochre (shaded with the same) and sepia; for the globe, yellow ochre and brown madder, orange and purple, shaded with sepia. The great division in the rings is coloured sepia" (not black as commonly described). "The pole and the narrow belts situated near it on the globe are pale cobalt blue." "These tints," he adds, are the nearest I could find to those seen on the planet; but there is a muddiness about all terrestrial colours when compared with the colours of the objects seen in the heavens. These colours could not be represented in all their brilliancy and purity, unless we could dip our pencil in a rainbow, and transfer the prismatic tints to our paper."

I can corroborate these remarks from observations made upon the planet with an $8\frac{1}{2}$-inch reflector. It is, indeed, a circumstance worthy of note, that the colours of the planets are much more strikingly exhibited by reflecting telescopes than by refractors, insomuch that, while Sir W. Herschel and Messrs. De la Rue and Lassell, making use of the former class of instruments, have all recorded the marked impression which the colours of Saturn and Jupiter have made upon them, we find that few corresponding observations have been made by observers who have been armed with even the most perfect specimens of the refracting telescope.

It must be noticed, however, that the colours of Saturn and his ring-system can only be seen in the most favourable observing weather.

XII.

FANCIED FIGURES AMONG THE STARS.

I THINK that every thoughtful student of the stars must have wondered how the figures of the various objects now pictured in our star-maps came to be imagined in the heavens themselves. It is a convenient answer to inquiries of the sort to say that it became necessary at an early stage in the progress of astronomy to have some means of identifying and naming star-groups, and that the arrangement into constellations was as suitable as any other that could have been desired. But it seems to me altogether unlikely that, in the infancy of a science, a mere arbitrary arrangement, such as this explanation supposes, should have been adopted. If we try to imagine the position of the first observers of the stars, what they wanted, and what they were likely to do,—and this *a priori* method of dealing with such questions is, I believe, the only safe one,—we perceive that the division of the stars into sets named after animals and other objects, without any real resemblance to suggest such nomenclature, is as unlikely a course as could possibly be conceived. Beyond all question, I think, the first watchers of the skies (they can scarcely be called astronomers) would have taken advantage of imagined similarity, more or less close, between each remarkable group of stars and some known object, to identify the group, and to obtain a name by which to speak of it.

Yet it must be admitted that, as the constellations are at present arranged and figured, it is very difficult, in the great majority of cases, to imagine the least resemblance between a constellation and the object from which it derives its name. This is not only true of the modern constellations, the preposterous *pneumatic machines, printing presses, microscopes,* and so forth, with which Hevelius and his successors foolishly crowded the heavens. Even the oldest of the old constellations of Ptolemy, nay, some even of those which are found among all nations, present, according to their present configuration, scarce any resemblance to their antitypes. For instance, it is well known that the Great Bear was recognised by many nations besides the Greeks and those, whoever they may have been, from whom the Greeks derived the constellation. We learn that when America was discovered the Iroquois Indians called this constellation Okouari, or the Bear. So the inhabitants of Northern Asia, the Phœnicians, the Persians, and others, called this constellation the Bear. The Egyptians, not knowing the bear, called the constellation the Hippopotamus, an animal resembling the bear in several respects, as in its heavy body, short inconspicuous tail, small head, and short ears. Yet the constellation, as at present figured, is certainly not in the remotest degree like a bear. Apart from the enormous tail given in the pictures to the bear (almost tailless in reality), it is impossible for the liveliest imagination to recognise a bear as the constellation is at present formed. Flammarion says that, "even if we take in the smaller stars that stand in the feet and head, no ingenuity can make it in this or any other way resemble a bear," adding the absurd explanation given by Aristotle, "that the name is derived from the fact that of all human animals the bear was thought to be the only one that dared to venture into the frozen regions of the north, and tempt their solitude and cold." As though the shepherds and tillers of the soil, who first gave names to the stars, were likely to consider such far-fetched reasons, even if they had known either the habits of the polar bears or had considered the relation of the northern star-groups to the polar regions of the earth.

Now the question whether any real resemblance

attracted the attention of the earlier observers in such cases as this is by no means without interest. If such a resemblance formerly existed, and does not now exist, it would follow that quite a considerable proportion of the stars have changed in brightness. Considering that each star is a sun, the centre, most probably, of a system like that which circles around our own sun, such a conclusion would be very startling indeed. It would have a special interest for ourselves, somewhat in the same way that the news that many railway accidents occur has an interest for those who travel much by rail. If accidents frequently happen to those other suns, in such sort that they either lose or gain greatly in brightness, an accident of one or other kind might well happen to our own sun, in which case the inhabitants of this earth would perish. For many of the stars, by our supposition, would have changed so much as either to lose their character as the defining stars of a constellation or by accession of brightness to acquire that character which in old times they had not possessed. Now, assuredly, a change of brightness competent to affect *our* sun's character (as viewed from any remote star system) in equal degree, would be destructive to the inhabitants of the earth. None at least of the higher races of animals or plants could bear the intense cold resulting from a change of the former kind, or the intense heat resulting from a change of the latter kind. Yet, if the constellations were once named because of their imagined resemblance to various objects, and if no such resemblance can now be even imagined, a change of one or other kind in the condition of our sun must be regarded as probable,—much in the same way that a regular traveller by train on any line must be regarded as exposed to danger, if accidents are known to be continually happening on that line.

What I now propose to do is to inquire whether we may not find the true figures and proportions of the ancient constellations in another way—viz., not by looking for them among the constellations as at present bounded and figured in our star-maps, but by searching the heavens themselves for them. This general method of search occurred to me very long ago while I was preparing various star-atlases, but the special mode of illustration here adopted occurred to me lately, while preparing for young astronomers in the United States a series of monthly maps showing the skies towards the north, south, east, and west, at different times of the night all the year round, and in various latitudes within the limits of the States. When I was in America I noticed, as I travelled about over a tolerably wide range of latitude, that the varying attitudes assumed by several of the constellations suggested features of resemblance to different objects. In constructing maps, simple in appearance, but based in reality on careful calculations, this characteristic came out more clearly. Adopting a particular way of presenting the connection between the various stars of a constellation, I often found the figure suggested which had actually been associated with the group of stars thus connected. Lastly, the idea of extending this method to other cases naturally occurred to me, and some of the results are presented in the present essay.

The method of delineation referred to is simply that of connecting the stars of a group by lines, *ad libitum*, that is, not merely introducing so many lines as will connect all the stars into a single set, but where necessary to complete the delineation of the imagined figure, adding other lines connecting pairs of stars belonging to the group, yet not so many that *every* pair of stars is connected by a line. The lines, again, need not be straight. On the contrary, where a group of stars forms a stream, the natural way of joining them is by lines so curved as to follow the serpentine course thus suggested. And in other cases a slight curvature of the lines joining pairs of stars will seem permissible, because corresponding to a configuration suggested by the stars themselves.

It is easily seen that in some of the simplest cases, the figure associated with a constellation is at once suggested by this method of delineation. For instance, take the case of the Northern Crown.

In this constellation we have a group which, while consisting of only a few stars, yet suggests very naturally the idea of a coronet of gems, as shown in fig. 32. The same is true also, though perhaps in less degree, of the Dolphin, as shown in fig. 33. It is noteworthy, by the way, that this constellation can hardly have been invented by landsmen. For though in our own time when the pictures of sea-creatures are accurately drawn, so that

Fig. 32.—The Northern Crown.

Fig. 33.—The Dolphin.

persons who have never been to sea may have a correct idea of the figure of such creatures, in old times it was exceedingly unlikely that any but sailors would have such familiar knowledge of the dolphin as to be reminded of that creature by a group of stars.

A much more complex constellation than either of those just mentioned—the Scorpion,—is even better represented by lineation, as shown in fig 34. It is not, however, with cases so remarkable as these that the difficulty suggested at the outset is really connected. The instances of really remarkable resemblance are so few that they must be regarded as altogether exceptional. The best proof that the Scorpion is unmistakably pictured by the stars is to be found in the fact that the

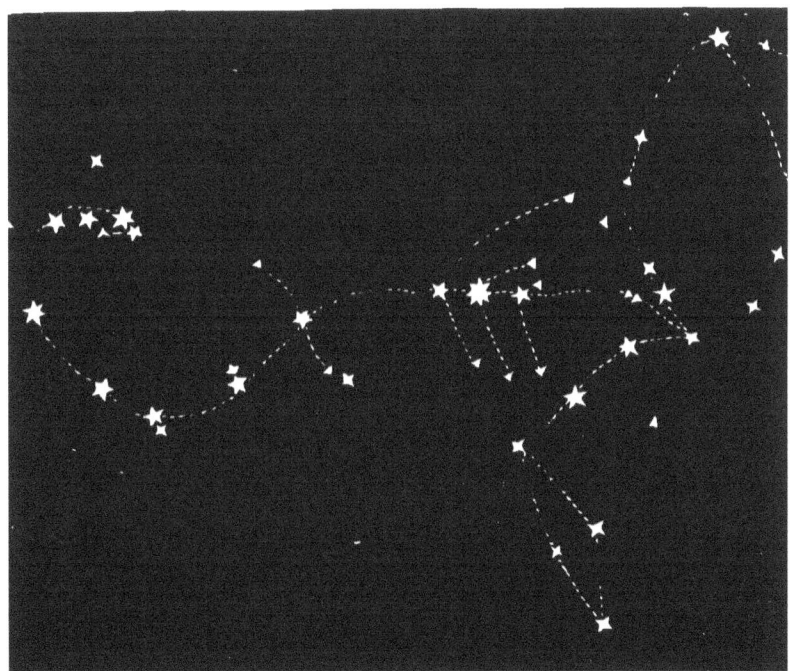

Fig. 34.— The Scorpion.

modern map-makers have not in this case departed much from the older delineations. No one, in fact, who knows what a Scorpion is like, could have any doubt as to the configuration of the body, at least, of the celestial Scorpion. So that though such a case illustrates well the way in which the method of delineation I have suggested may be made to picture the object seen by the ancient observers in the heavens, it does not afford any answer to the difficulty indicated by those who assert that the Great Bear, the Lion, the Ship, and other of the old constellation figures, have no real existence among the stars.

Before leaving the Scorpion, however, I must call attention to one or two points which this remarkable constellation seems to establish. First, it is clear that in its case real resemblance suggested the association of a group of stars with a familiar object. Since this resemblance remains, we infer that the group of stars presents now an appearance closely resembling that which it presented four or five thousand years ago. And as there is no special reason why the stars of the Scorpion more than those of other constellations should retain their lustre unchanged, we gain a certain probability for the belief that *all* the constellations are now very much as they were when first named. Indeed, it so happens that the region occupied by the Scorpion is perhaps that part of the heavens where changes would on the whole most probably occur, the region of the Milky Way crossed by the Scorpion being exceptionally irregular. We may note also that the part of the earth where the observers lived who called this constellation the Scorpion must have been one where the reptile is well known, a conclusion which seems to dispose of the belief that the first astronomers lived in high latitudes.

Let us, now, however, take some of the more difficult cases. We cannot do better, perhaps, than take at the very outset the Great Bear, a constellation of which many astronomers have asserted that it no longer presents and probably never did present the slightest resemblance to a bear.

I would lay down, in the first place, the hypothesis that the stars in the region of the heavens now occupied by the Great Bear *must* have reminded the earliest observers of a large, heavily-bodied, small-headed, short-eared, and short-tailed creature, such as either a bear or a hippopotamus. Next, it may be taken for granted that the creature of which they were thus reminded was one with which they were familiar; and as we have already seen that the inventors of the oldest constellations cannot have lived in very high latitudes, we may conclude with great probability that the bear imagined in the heavens was not the Polar bear, but the bear from which the first shepherd astronomers had to defend their herds and flocks, —the Syrian bear, as it is commonly called, though the species inhabited also the greater part of Asia Minor in former times. The Indians may be supposed to have seen the grizzly bear, not the smaller black bear, in the heavens. The features to be looked for, then, among the stars, are those common to the bears of comparatively low latitudes—not those of the polar bear.

So much premised we may proceed to inquire whether the region of the heavens occupied by the Great Bear presents such a creature with sufficient distinctness to suggest the idea of the animal to persons familiar with its aspect.

It is perhaps hardly necessary to remark that we must not expect to find a complete far less a perfect picture of a bear, or lion, or ship, in a large region of the heavens such as is occupied by these constellations. If some characteristic feature of a bear could be recognised in a group of stars, the ancient observer would be content to recognise the region of the heavens which would be occupied by the entire figure of the animal, as belonging to a Great Bear, unless some marked peculiarity in the stars of that region absolutely prevented the most lively imagination from conceiving a bear's body there. As an instance of the latter kind may be mentioned the Bull and the Ship, both of which constellation figures are seen only in part. The Bull's head is exceedingly well marked, as is the stern of the ship Argo, but the liveliest imagination cannot recognise the body and tail of a bull, or the fore-part of a ship, where these should be. Consequently the ancients always regarded the Bull as a half bull,

Fig. 35.— The Great Bear.

and (as Aratus is careful to mention) they recognised only the stern of the good ship Argo. But in general, where only some marked feature of an object could be imagined, or perhaps two or three, they yet conceived the whole object to be shown in the heavens, though it may have been altogether impossible to distribute the other stars over the remaining portion of the object in such a way as to show any natural association.

The Great Bear seems to have been a constellation of this sort. One can recognise the head of an animal like the bear or the hippopotamus, and also the feet of such a creature, but the proper disposal of the stars forming the animal's body is not so easy. This would not interfere, however, with the choice of the bear to represent the region of stars occupied by the constellation. Every one who has seen faces and figures in the fire—and who has not?—knows that one or two features will suffice to suggest a resemblance; either the imagination does all the rest, or else the idea is suggested that some other object partially conceals that portion of the imagined figure which is wanting.

Fig. 35 shows how, as I conceive, a bear was figured in the heavens by those who, in various nations, gave to the stars of this part of the sky the name of the Great Bear.

It will be noticed in the first place that the famous Septentriones (the seven stars of the Plough, as in England the set is called, the Dipper as it is called in America, the Corn-measurer as it was called by the ancient Chinese) has little or nothing to do with the configuration of the Bear, though forming a part of the constellation. It is the set of small stars forming the head which seems to have suggested the idea of a bear, though two of the paws are also well defined by the stars. But the outlining of the head of a bear or hippopotamus is really sufficiently close to require no very lively imagination to fill it in. Fig. 36, giving these stars only, serves to show this, I think. That the entire figure of a bear or hippopotamus was not recognised seems further shown by the figure assigned to the constellation in the Zodiac of Tentyra, or Denderah, where it appears as in fig. 37. The smaller figure is supposed to represent the Little Bear.

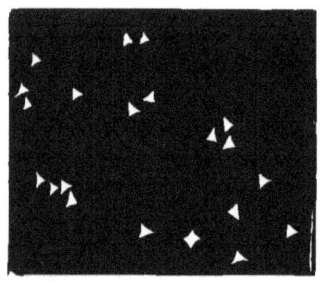

Fig. 36.—The Bear's Head.

In the second place, the reader familiar with the constellations will perceive that several stars not at present appertaining to the Great Bear are included within the configuration itself of the animal in fig. 35. Thus the third magnitude star behind the right ear belongs to

FANCIED FIGURES AMONG THE STARS. 251

the constellation of the Dragon; the third magnitude star near the hind quarters is Cor Caroli, the chief star of the modern constellation *Canes Venatici*, or the Hunting Dogs. It appears to me that we ought not to expect that the first observers of the heavens, in recognising

Fig. 37.—The constellations of the Bears, represented as a hippopotamus (?) and wolf (?) in the Denderah Zodiac.

imaginary features of resemblance between a group of stars and some known object, would be careful to inquire whether some among those stars were included in a group which they had compared or might afterwards compare with another object. It is very necessary for the astronomer of our time, nay, it may have even been very necessary for the astronomers of the times of Hipparchus, Ptolemy, etc., to have the limits of the constellations clearly defined, and to let no conspicuous star be common to different constellations. But as regards the figures fancied in the heavens by the first observers of the stars, considerations of that sort would be of no importance whatever. Indeed, it is worthy of notice that even so late as the time of Bayer, who gave to the stars their Greek letters, the constellations were not separated from each other. He called the star now known as Beta Tauri only, Gamma Auriga *also*, so that now Auriga has stars Alpha, Beta, Delta, and so forth, but no Gamma. Similarly, we look in vain for any star Delta in the constellation Pegasus, simply because Bayer called one and the same star Alpha Andromedæ and Delta Pegasi, the astronomers of our own time retaining only the former name for this star,—the bright one adorning the head of Andromeda. Even in our time it has been found impossible properly to separate the older constellations from each other, so that to this day the Scorpion remains entangled with the legs of Ophiuchus, who is further inextricably mixed up with the Serpent. In fact, the Serpent is divided into two separate parts by the body of Ophiuchus, mapmakers having no choice but either to allow Ophiu-

Fig. 38.—The Original Constellation of the Lion.

chus to divide the Serpent, or the Serpent to divide Ophiuchus.

In the next case, that of the Great Lion, we have still further to depart from the modern configuration of the constellation. No one can imagine the remotest resemblance between any part of a lion and the grouping of stars falling on the corresponding portion of Leo in the modern constellation. The nose of the Lion now falls near λ (fig. 38); μ and ρ forming the outline of the mane, β the end of the tail, ϵ the nearer fore-paw, τ the nearer hind-paw. The original Lion, I cannot doubt, was imagined somewhat as pictured in fig. 38. The head and mane are unmistakably pictured among the stars, the paws fairly, the relatively small quarters and the tufted tail exceedingly well—always remembering that anything like very close resemblance is not to be looked for between a widely extended group of stars and the figure of an animal or other large object. If we remember also that uncultured nations, like children, are much quicker in imagining resemblances than those carefully trained to recognise the artistic delineation of objects, we cannot be surprised to find that nearly all those nations who were acquainted with the lion imagined a large leonine figure in the part of the heavens now centrally occupied by our modern and most puny Lion, but including portions of Cancer, the whole of Leo

Minor (one of Hevelius's absurd inventions), the Hair of Berenice, and a star or two belonging to Virgo.

We have to treat in a similar way the constellation

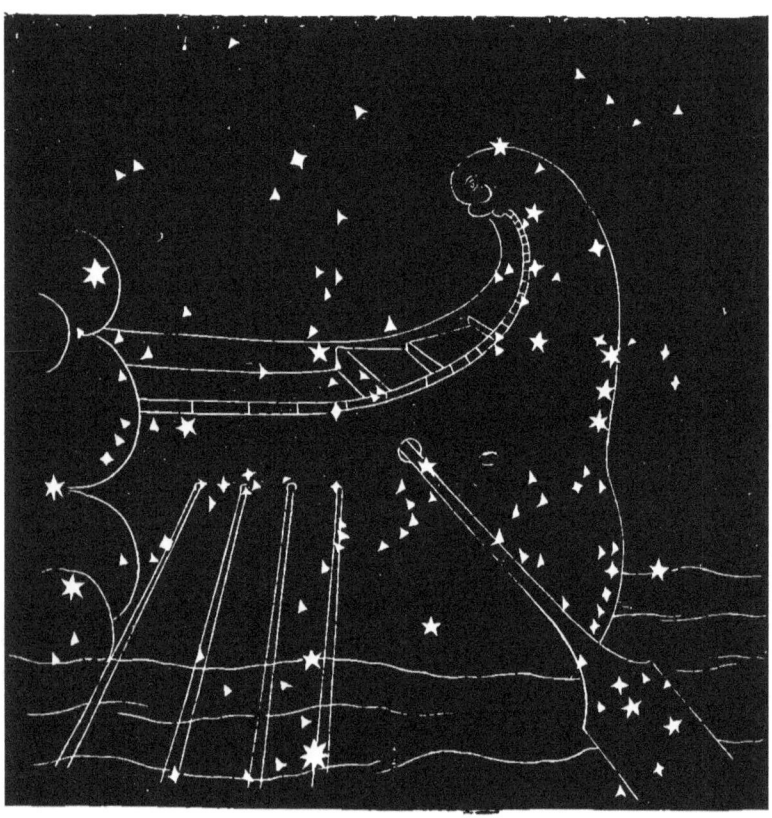

Fig. 39 — The Original Ship "Argo."

Argo of our present maps, to get the good ship Argo, as the ancients must have conceived the constellation. Fig. 39 shows the Ship as I imagine she was originally pictured. The stars which mark her curved poop belong in part at present (as doubtless they have long belonged) to the Larger Dog, while those which mark the steering-oar belong to the modern constellation Columba Noachi, or Noah's Dove. It must be observed that the bright star Canopus, shown in the water, was not visible in the time of the first observers in the latitude where they probably dwelt. The mighty gyrating motion of the earth has caused these stars to be brought five or six degrees further from the southern pole of the heavens. But Canopus and a few of the small stars near it are the only stars which have thus been added to the constellation as seen from the regions inhabited by the first observers. (Canopus was known to the Arabian and Egyptian astronomers.)

This introduces another point which seems worth noticing. At present the ship Argo is never seen from any part of the earth's surface as pictured in fig. 39. When due south, the position whence in all northern latitudes the constellation is most favourably seen, the ship is always tilted up at the stern: one would say, in more nautical phrase, she is down by the head, if the ship had any fore-part; but from time immemorial she has been a half-ship only. Some 4,000 years ago, however, Argo stood nearly on an even keel when due south. Again, it is to the mighty gyrational motion of the earth that we have to look for the cause of the great change in the apparent position of the ship. The sphere of the fixed stars has remained all the time unchanged, or very nearly so, but the direction in which the earth's axis of rotation points has swayed round (much as the axis of a reeling top sways round) through about one-sixth part of a complete gyration.

In the regions where astronomy first began as a science, Argo not only stood on an even keel but almost on the horizon when due south; and the features of resemblance to a ship, which I have endeavoured to portray in fig. 39, must have seemed much more striking there (and then) than now.

The fore-part of the ship, or rather that region of the heavens where the fore-part should be, is occupied by great masses of the Milky Way in one of its brightest and most remarkable portions. I have sometimes fancied that in some of the old Zodiac temples of star-worshippers the constellation Argo was depicted as a mighty ship, gemmed with stars, and heavily laden in its fore-part with great masses of gilded cloud to represent the Milky Way, and that from such representations of the constellation came the tradition of the ship Argo and its cargo of golden fleece. Many parts of the story of Jason and his companions seem to relate to objects depicted in the old constellation-domes,—as those relating to the Dragon, to Hercules, Castor and Pollux, the

Centaur, etc. There is also a curious reference, in the tradition, to the stern of the ship, which is much like what we can imagine as resulting from an attempt to explain the appearance of this part only, in the set of constellation figures. We read that the entrance to the Euxine Sea was fabled to be closed up by certain rocks called Symplegades (the Clashers), which floated on the water, and when anything attempted to pass through came together with such velocity that not even birds could escape. Phineas advised them to let a bird fly through, and if the bird passed safely, to venture the passage. It passed with only the loss of its tail; and the Argo, favoured by Juno, and impelled by the utmost efforts of its heroic crew, passed also, though so narrowly that the meeting rocks carried away part of her stern-works, which remained fixed there thenceforward.

For my own part, I think we may not only regard the story of the ship Argo as in reality a version, though much modified, of the account of Noah's deluge, but consider the series of constellations, Aquarius, Cetus, Eridanus, Argo, Corvus, Centaurus, Ara, and Sagittarius, as typifying the same narrative. It is somewhat curious that if we place these constellations in their original position,—that is, as they were before the changes which the earth's great gyration has introduced during the last four thousand years or so,—we find the following coincidences with the account of the deluge. First comes Aquarius (whose beginning would correspond with the sun's position on or about the seventeenth day of the second month of the old Pleiades year) pouring water. His range on the ecliptic (or the space he occupies in the annual range represented in the zodiac temple) is about forty days. Then came the watery constellations Eridanus, the river, and Cetus, the sea monster, having, with the ship Argo, a range of about 150 days of the annual circuit. About forty days later in the circuit we find Corvus, the raven, whose feet rest on Hydra, the great celestial sea-serpent, as though no dry land could be found by the bird. A dove also, if we accept the interpretation above given of the Argo narrative, may have been represented in this part of the star temple. Next we have the Centaur, originally we know represented as a man only, offering an animal as sacrifice on the altar Ara. There is a cloud of stars rising from the altar: we may recall Manilius's account of the constellation,—

"Ara, ferens thuris, stellis imitantibus, ignem." *

---

* "The altar, bearing fire of incense, pictured by stars." A remarkably bright and complex portion of the Milky Way lies near the constellation Ara, giving the appearance of smoke ascending from the altar, only the altar must be set upright, as in my Gnomonic Atlas, not inverted as in all the modern maps. (It is shown properly in the old Farnese globe).

In this cloud is the Bow of Sagittarius, the bow being originally alone shown, as it is indeed the only figure which can be imagined among the stars of this region. So that these constellation figures seem to typify Noah offering sacrifice on the Altar, and the Bow of Promise set in the cloud above the altar. It is curious, too, that while the time of Noah's leaving the ark was a year and ten days from the beginning of the rains, the constellation Sagittarius overlaps the conjoined watery signs Capricornus and Aquarius (running south of them) by about so much as would correspond to ten days of the annual circuit of the heavens.

The objections to the view of matters above indicated are, first, that the constellations referred to seem to have been formed because of real resemblance between the star-groups and the figures associated with them; and, secondly, that the Zodiac temples were probably erected by star-worshippers, and would scarcely have been employed to typify such a narrative as that of the Deluge. The theory that the narrative itself was an attempt to interpret pictures represented on a Zodiac temple will, of course, be objectionable to many readers; though they may not be unwilling to believe that the fable of the Argonautic expedition had its origin in some such way.

It will have been noticed that in the figures which I

have given of the Great Bear, Lion, and Ship, I have not altogether adhered to my idea of simply connecting the stars of a group by lines. To say the truth, although a rough notion of a bear, lion, or ship may thus be given, the figure so presented is not altogether satisfactory to the mind. In any case, as for instance even in the Scorpion (of all these figures the best marked), the line-figure is very imperfect. But in some cases it does suggest the idea of an animal or figure, or a part of either, much in the same way that the idea of a human figure can be suggested by a few lines forming a skeleton figure, such as our old friend Tommy Traddles used to draw. Now the Lion, Bear, and Ship are not well suited for this sort of delineation, as anyone will find who tries to suggest the idea of a bear, lion, or ship (of the old-fashioned heavily-sterned sort) by means of a few lines.

In order, however, to show that in some cases a skeleton figure can be formed by joining the stars of a constellation, and that the figure thus formed represents (of course in an utterly inartistic sort of way) the object associated with those stars, I will now take one or two instances in which such resemblance suggested itself to me without being specially sought for. I might add to the Crown, Dolphin, and Scorpion, the Chair of Cassiopeia, the figure of Orion, and the constellation of the Cup; I omit these, however, not because they are unfit for my purpose, but because they so obviously illustrate my argument. No one, with the least power of imagination, can fail to see how a chair, a belted giant, and a cup, are pictured, as it were, in these constellations. I will take others where the resemblance is less obvious.

Thus, I think scarcely anyone who is acquainted with the constellation Andromeda can have failed to be perplexed by the association of the figure of a chained lady with this group of stars. In the arrangement of the stars themselves, without lines drawn to connect them, no such figure can be imagined; at least I fail utterly for my own part when I attempt to picture such a figure, even now that I recognise how the figure is formed, skeleton-wise, by connecting lines. I cannot but think this figure *must* have been imagined from pictures of the groups of stars with lines connecting them, and not from the stars themselves. There is this reason, among others, for so thinking, The lady's head is represented by a single star, Alpherat. Now a single star in the sky, however bright, is not large enough to represent the head of a human figure like Andromeda's. But the representation of a bright star like Alpherat in a chart or sculpture has sufficient size to serve for a head, because size is the only way in which brightness can be indicated.

In fig. 40 the stars forming the constellation Andromeda are shown; also the chair of Cassiopeia; and, on the right, one of the fishes and the triangle. A group of stars in the upper left-hand corner marks the place of

Fig. 40.—Andromeda.

the rock to which the chains are fastened which bind Andromeda's right hand.

It cannot be said that the skeleton picture shewn in fig. 40 is very graceful or artistic; but, on the other hand, it cannot, I think, be doubted that there is enough in it to suggest the idea of a chained person. The fish naturally suggests the idea that the place is by the sea-shore. And the chair suggests the idea of some one on the shore waiting and watching. In our own time, probably, the idea suggested would be that of a person taking a bath, while some one sat in a chair on the sands and waited for their turn. But to the old observers of the heavens, unfamiliar as they were with sea-side diversions, the notion would more naturally occur of a woman chained to a rock,

Lifting her long white arms, widespread, to the walls of the basalt;

while not far off was imagined among the stars the monster Cetus coming onward,

> bulky and black as a galley,
> Lazily coasting along, as the fish fled leaping before it.

One of these fish is seen close by the figure of the chained Andromeda. Near at hand they imagined the father and mother of the lady; Cassiopeia sitting close to the shore; but

> Cepheus far in the palace
> Sat in the midst of his hall, on his throne, like a shepherd of people,
> Choking his woe dry-eyed, while the slaves wailed loudly around him.

The story of Andromeda, as the reader doubtless knows, is not of Greek origin. Its real origin is lost in a far antiquity. The Indians have the same story in their astronomical mythology, and almost the same names. Thus Wilford, in his Asiatic Researches, relating his conversation with an Indian astronomer, says, " I asked him to show me in the heavens the constellation of Antarmada, and he immediately pointed to Andromeda, though I had not given him any information about it beforehand. He afterwards brought me a very rare and curious work in Sanscrit, which contained a chapter devoted to *Upanachatras*, or extra-zodiacal constellations, with drawings of *Capuja* (Cepheus), and of *Casyapi* (Cassiopeia), seated and holding a lotus flower in her hand ; of *Antarmada*, chained, with the fish beside her ; and last, of *Parasiea* (Perseus), who, according to the explanation of the book, held the head of a monster which he had slain in combat ; blood was dropping from it, and for hair it had snakes."

As another illustration of the method I have described, I give the constellation Pegasus, or, as it was sometimes called, the Half-horse. I do not assert that fig. 41 presents a very well shaped steed, any more than that in fig. 40 a lady of exquisite proportions is pictured. But one can perceive how the stars suggest the idea of a horse in one case, and of a human figure with upraised fastened arms in the other. It is commonly stated that Pegasus is one of the constellations showing no resemblance at all to the figure associated with it. I think fig. 41 suffices to show that there is some slight resemblance at least.

It may be mentioned, in passing, that all the nations of antiquity would not be likely to form equally clear

Fig. 41.—Pegasus.

conceptions of figures in the heavens. There are marked differences between the various races of the human family in this respect, just as there are marked differences between various persons in the power of imagining figures under different conditions. Some persons see figures at once in a cloud, in the outline of a tree, in a fire, in a group of accidental markings, and so forth; while others not only do not see such figures, but cannot imagine them even when their outlines are indicated. So it is with different races of men. There have been some which, even when only just emerging from the utterly savage state, possessed so much of the imaginative power as to be able to picture for themselves, by lines cut with rude flint instruments on pieces of bone, horn, or ivory, the animals with which they were familiar. We have even among such pictures some belonging to an age so remote that the mammoth (or hairy elephant) had not yet entirely disappeared from Europe; for, in the cave of La Madeleine, at Dordogne, among other relics of the stone age, there has actually been found a drawing of the mammoth scratched on a piece of mammoth tusk. On the other hand, there are some races in existence at the present day, in a more advanced stage of civilization, who cannot perceive even in well-executed coloured drawings any resemblance to the objects pictured. An aboriginal New Hollander, says Oldfield, "being shown a coloured engraving" of a member of his own tribe, "declared it to be a ship, another a kangaroo, and so on; not one of a dozen identifying the portrait as having any connection with himself." A rude drawing, with all the lesser parts much exaggerated, they can realise. Thus, to give them an idea of a man, the head must be drawn disproportionately large. Dr. Collingwood tells us that when he showed a copy of the *Illustrated London News* to the Kibalaus of Formosa, he found it impossible to interest them by pointing out the most striking illustrations, "which they did not appear to comprehend." Denham (I quote throughout from Lubbock's most valuable and interesting work on the Origin of Civilization) says that Bookhaloum, a man otherwise of considerable intelligence, though he readily recognised figures, could not understand a landscape. "I could not," he says, "make him understand the print of the sand-wind in the desert, which is really so well described by Captain Lyons' drawing. He would look at it upside down; and when I twice reversed it for him he exclaimed, 'Why! why! it's all the same.' A camel or a human figure was all I could make him understand, and at these he was all agitation and delight. 'Gieb! Gieb!—wonderful! wonderful!' The eyes first took his attention, then the other features; at the sight of the sword, he exclaimed, 'Allah! Allah!' and, on discovering the guns, instantly exclaimed, 'Where is the powder?'"

We have in the consideration of this diversity of character between different races and nations, as respects the power as well of imagining as of delineating figures (the two are closely connected), one means of judging to what race we owe the original constellations. For although some figures in the heavens are manifest enough, others require a considerable power of imagination. And it should be noted that this must have been true even if we suppose (which I think I have succeeded in showing we need not do) that many of the stars have changed in brightness, and that thus resemblances have disappeared which formerly existed. For, in any case, the heavens four, ten, or twenty thousand years ago, or at whatever remote period we set the original invention of the constellations, must have presented the same characteristics as at present. It can never have been the case that all the star-groups could be compared at once, obviously, with the figures of men and animals. So that only a race of lively imagination could have found figures for all the star-groups, as was certainly done in very remote times by some race.

The race, then, to whom we owe the general system of constellations, was probably one with so much talent for artistic delineation that in later ages this people would have become distinguished for skill in painting and sculpture. I think the sculptures found in Babylon, and the traditions left of the artistic skill of the Babylonians, correspond well with the belief that the constellations had their origin, and astronomy its first development, among that people or a kindred race.

But the chief lesson to be derived (and I think it may

fairly be derived) from the study of the constellation-groups is, that enough resemblance still remains, if only the arbitrary boundaries invented for the constellation figures in recent times are overlooked, to assure us that no very great changes have taken place in the aspect of the heavens for thousands of years. A few stars here and there have certainly changed greatly in brightness, and some few have changed considerably even in position; while a considerable number have probably changed slightly in brightness, and all, or very nearly all, have changed somewhat in position. But on the whole the aspect of the stellar heavens now is the same as it was when the constellation figures were first imagined.

This thought not only assures us of the permanence of our own sun (seeing that among the thousands of his fellow-suns which spangle the heavens so few have changed in lustre), but seems to me to give to the study of the stars a singular charm. Our antiquaries and archæologists present for our study the relics of long past ages, and we may often rest assured that the objects thus gathered for us were really used in old times, though probably in a manner not understood by us, and when in a condition very unlike that in which they have reached our times. In nearly all such instances, however, doubt exists as to the antiquity of the relic, as to the race to whom it really belonged, and as to its real use and purport. But as regards the stellar heavens we have no doubt. Of all the objects on which the eyes of remote races have rested, the celestial bodies are undoubtedly the most ancient, while at the same time they and they alone were most certainly contemplated by all mankind. From the very earliest ages, from the time when the child-man first turned his thoughts from mere animal wants to the wonders of nature, the stars, and the sun and moon and planets must have drawn to themselves the attention of all who had eyes to see even though they had no power to understand the glories of the star-depths. Men pictured among the stars the objects most familiar to them, the herds and flocks which they tended, the herdsman himself, the waggoner, the huntsman, the birds of the air, the beasts of the field, the fishes of the sea, the ship, the altar, the bow, the arrow, and, one may say, all that according to their knowledge existed in the heavens above, in the earth beneath, and in the waters under the earth. Imperfect and anomalous as these meanings are, in relation to modern astronomy, with its exact methods, elaborate instruments, and profound investigations into the meaning of all the phenomena of the heavens, they nevertheless retain their place, and are likely long to do so, in virtue of the hold which they took, in remote ages, on the imagination of mankind in general.

XIII.

TRANSITS OF VENUS.

AS a transit of Venus, visible in this country, occurs in December, 1882, my readers, although they may not care for an account of the mathematical relations involved in the observation and calculation of a transit, will probably be interested by a simple explanation of the reasons why transits of Venus are so important in astronomy.

Of course it is known that a transit of Venus is the apparent passage of the planet across the face of the sun, when, in passing between the earth and sun, as she does about eight times in thirteen years, she chances to come so close to the imaginary line joining the centres of those bodies that, as seen from the earth, she appears to be upon the face of the sun. We may compare her to a dove circling round a dovecot, and coming once in each circuit between an observer and her house. If in her circuit she flew now higher, now lower, or, in other words, if the plane of her path were somewhat aslant, she would appear to pass sometimes above the cot, and sometimes below it, but from time to time she would seem to fly right across it. So Venus, in circuiting round the sun, appears sometimes, when she comes between us and the sun, to pass above his face, and sometimes to pass below it; but occasionally passes right across it. In such a case she is said to transit the sun's disc, and the phenomenon is called a transit of Venus. She has a companion in these circuiting motions, the planet Mercury, though this planet travels much nearer to the sun. It is as though, while a dove were flying around a dovecot at a distance of several yards, a sparrow were circling round the cot at a little more than half the distance, flying a good deal more quickly. It will be understood that Mercury also crosses the face of the sun from time to time—in fact, a great deal oftener than Venus; but, for a reason presently to be explained, the transits of Mercury are of no great importance in astronomy. One occurred in 1861, another in 1868; another in May, 1878; yet very little attention was paid to those events; and before the next transit of Venus, in 1882, there will be a transit of Mercury, in November, 1881; yet no arrangements have been made for observing Mercury in transit on these occasions;

whereas astronomers began to lay their plans for observing the transit of Venus in 1882, as far back as 1857.

The illustration which I have already used will serve excellently to show the general principles on which the value of a transit of Venus depends; and as, for some inscrutable reasons, any statement in which Venus, the sun, and the earth are introduced, seems by many to be regarded as, of its very nature, too perplexing for anyone but the astronomer even to attempt to understand, my talk in the next few paragraphs shall be about a dove, a dovecot, and a window, whereby, perhaps, some may be tempted to master the essential points of the astronomical question who would be driven out of hearing if I spoke about planets and orbits, ascending nodes and descending nodes, ingress and egress, and contacts internal and external.

Suppose D, fig. 42, to be a dove flying between the window A B and the dovecot C c, and let us suppose that a person looking at the dove just over the bar A sees her apparently cross the cot at the level a, at the foot of one row of openings, while another person looking at the dove just over the bar B sees her cross the cot apparently at the level b, at the foot of the row of openings next above the row a. Now suppose that the observer does not know the distance or size of the cot, but that he does know in some way that the dove flies *just* midway between the window and the cot; then it is perfectly clear that the distance a b between the two rows of openings is exactly the same as the distance A B between the two window-bars; so that our observers need only measure A B with a foot-rule to know the scale on which the dovecot is made. If A B is one foot, for instance, then a b is also one foot; and if the dovecot has three equal divisions,

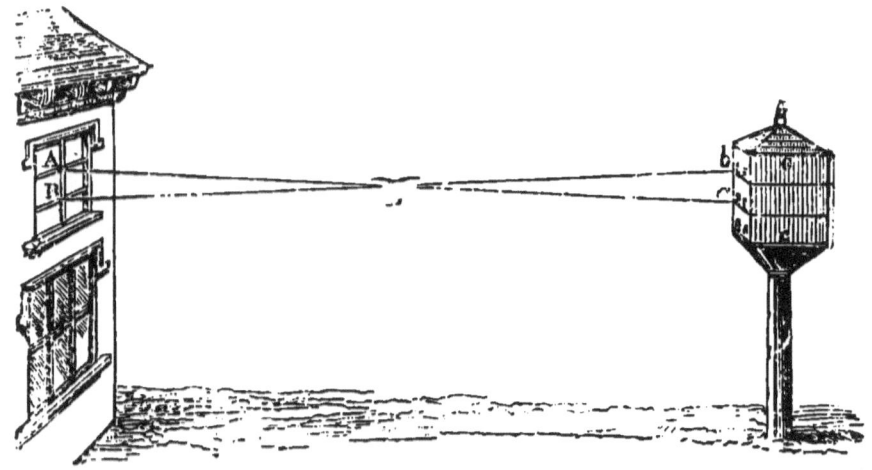

Fig. 42.

as shown at the side, then C c is exactly one yard in height.

Thus we have here a case where two observers, without leaving their window, can tell the size of a distant object.

And it is quite clear that wherever the dove may pass between the window and the house, the observers will oe equally able to determine the size of the cot, if only they know the relative distances of the dove and dovecot.

Thus, if D a is twice as great as D A, as in fig. 43,

Fig. 43.

then *a b* is twice as great as A B, the length which the observers know; and if D *a* is only equal to half D A, as in fig. 44, then *a b* is only equal to half the known

Fig. 44.

length A B. In every possible case the length of *a b* is known. Take one other case in which the proportion is not quite so simple:—Suppose that D *a* is greater

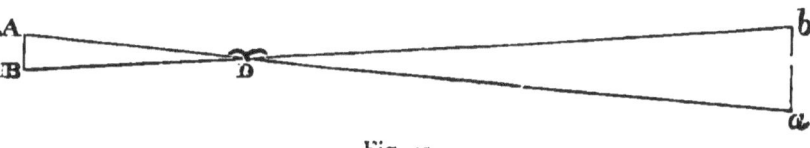

Fig. 45.

than D A in the proportion of 18 to 7, as in fig. 45; then *b a* is greater than A B in the same proportion; so that, for instance, if A B is a length of 7 inches, *b a* is a length of 18 inches.

We see from these simple cases how the actual size of a distant object can be learned by two observers who do not leave their room, so long only as they know the relative distances of that object and of another which comes between it and them. We need not specially concern ourselves by inquiring *how* they could determine this last point: it is enough that it might become known to them in many ways. To mention only one. Suppose the sun was shining so as to throw the shadow of the dove on a uniformly paved court between the house and the dovecot, then it is easy to conceive how the position of the shadow on the uniform paving would enable the observers to determine (by counting rows) the relative distances of dove and dovecot.

Now, Venus comes between the earth and sun precisely as the dove in fig. 45 comes between the window A B and the dovecot *b a*. The relative distances are known exactly, and have been known for hundreds of years. They were first learned by direct observation; Venus going round and round the sun, within the path of the earth, is seen now on one side (the eastern side) of the sun as an evening star, and now on the other side (the western side) as a morning star, and when she seems farthest away from the sun in direction E V (fig. 46) in one case, or E *v* in the other case, we know that the line E V or E *v*, as the case may be, must just touch her path; and perceiving how far her place in the heavens is from the sun's place at those times, we know, in fact, the size of either angle S E V or S E v, and, therefore, the shape of either triangle S E V or S E v. But this amounts to saying that we know what pro-

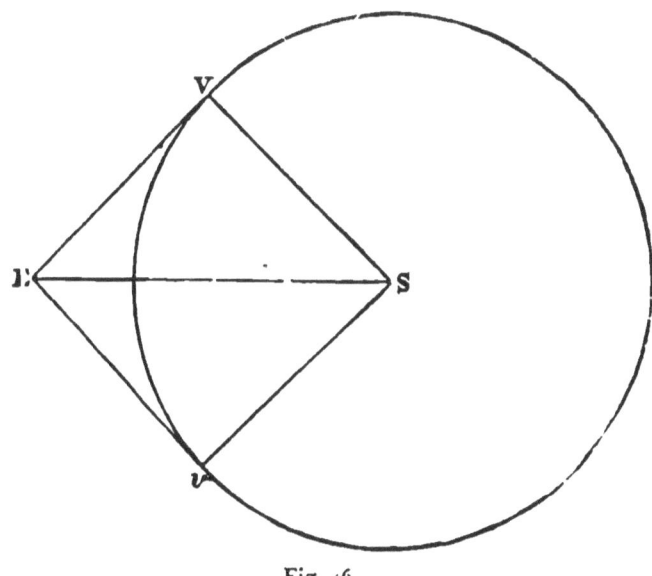

Fig. 46.

portion S E bears to S V—that is, what proportion the distance of the earth bears to the distance of Venus.*

---

* There is, however, a much more perfect way of determining this proportion, by applying the law which Kepler found to connect the distances of the planets from the sun with the times in which they complete the circuits of their orbits. The law is that, if we take any two planets, and write down the numbers expressing their periods of circuit (say in days), and the numbers expressing their distances from the sun (say in miles) in the same order; then if we multiply each number of the first pair into itself, and each

This proportion has been found to be very nearly that of 100 to 72; so that when Venus is on a line between the earth and sun, her distances from these two bodies are as 28 to 72, or as 7 to 18.

These distances are proportioned precisely then as D A to D *a* in fig. 45; and the very same reasoning which was true in the case of dove and dovecot is true when for the dove and dovecot we substitute Venus and the sun respectively, while for the two observers looking out from a window we substitute two observers

number of the second pair twice into itself, the four numbers thus obtained will be proportional; that is to say, as the first is to the second, so will the third be to the fourth. Now, as every one knows who has worked sums in the rule of three, when any three are given out of four proportionals, the fourth can always be found; but we know the periods of circuit both of the earth and Venus (365·2564 days and 224·7008 days respectively) very exactly indeed, because they have traversed their orbits so many times since they began to be observed by astronomers. We can call the earth's distance 100, and then applying the rule just stated, we get Venus' distance relatively to the earth's. The reader who cares to work out this little sum will find no difficulty whatever—if at least he is able to extract the cube roots of any number. The proportion runs thus :—

$$365\cdot2564 \times 365\cdot2564 : 224\cdot7008 \times 224\cdot7008$$
$$:: 100 \times 100 \times 100 : (\text{Venus' distance cubed.})$$

Work out this sum and we get for Venus' distance 72·333. The ratio of Venus' distance to the earth's is almost exactly expressed by the numbers 217 and 300.

stationed at two different parts of the earth. It makes no difference in the essential principles of the problem that in one case we have to deal with inches, and in the other with thousands of miles; just as in speaking of fig. 45 we reasoned that if A B, the distance between the eye-level of the two observers, is 7 inches, then *b a* is 18 inches, so we say that if two stations, A and B, fig. 47, on the earth E, are 7000 miles apart (measuring the distance in a straight line), and an observer at A

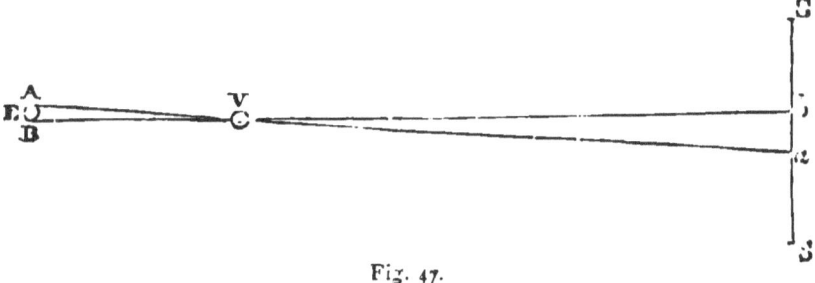

Fig. 47.

sees Venus' centre on the sun's disc at *a*, while an observer at B sees her centre on the sun's disc at *b*, then *b a* (measured in a straight line, and regarded as part of the upright diameter of the sun) is equal to 18,000 miles. So that if two observers, so placed, could observe Venus at the same instant, and note exactly where her centre seemed to fall, then since they would thus have learned what proportion *b a* is of the whole diameter S S' of the sun, they would know how many miles there are in that diameter. Suppose, for instance, they found, on comparing notes, that *b a* is about the 47th part of the whole diameter, they would know that the diameter of the sun is about 47 times 18,000 miles, or about 846,000 miles.

Now, finding the real size of an object like the sun, whose apparent size we can so easily measure, is the same thing as finding his distance. Any one can tell how many times its own diameter the sun is removed from us. Take a circular disc an inch in diameter, —a halfpenny, for instance—and see how far away it must be placed to exactly hide the sun. The distance will be found to be rather more than 107 inches, so that the sun, like the halfpenny which hides his face, must be rather more than 107 times his own diameter from us. But 107 times 846,000 miles amounts to 90,522,000 miles. This, therefore, if the imagined observations were correctly made, would be the sun's distance.

I shall next show how Halley and Delisle contrived two simple plans to avoid the manifest difficulty of carrying out in a direct manner the simultaneous observations just described, from stations thousands of miles apart.

We have seen that the determination of the sun's distance by observing Venus on the sun's face would be a matter of perfect simplicity if we could be quite sure that two observations were correctly made, and at exactly the same moment, by astronomers stationed one far to the north, the other far to the south.

The former would see Venus as at A, fig. 48, the other

Fig. 48.

would see her as at B; and the distance between the two lines $a\,a'$ and $b\,b'$, along which her centre is travelling, as watched by these two observers, is known quite certainly to be 18,000 miles, if the observers' stations are

7,000 miles apart in a north-and-south direction (measured in a straight line). Thence the diameter S S' of the sun is determined, because it is observed that the known distance $a\ b$ is such and such a part of it. And the real diameter in miles being known, the distance must be 107 times as great, because the sun looks as large as any globe would look which is removed to a distance exceeding its own diameter (great or small) 107 times.

But unfortunately it is no easy matter to get the distance $a\ b$, fig 48, determined in this simple manner. The distance 18,000 miles is known; but the difficulty is to determine what proportion the distance bears to the diameter of the sun S S'. All that we have heard about Halley's method and Delisle's method relates only to the contrivances devised by astronomers to get over this difficulty. It is manifest that the difficulty is very great.

For, first, the observers would be several thousand miles apart. How then are they to ensure that their observations shall be made simultaneously? Again, the distance $a\ b$ is really a very minute quantity, and a very slight mistake in observation would cause a very great mistake in the measurement of the sun's distance. Acordingly, Halley devised a plan by which one observer in the north (or as at A, fig. 47) would watch

Venus as she traversed the sun's face along a lower path, as $a\ a'$, fig. 49; while another in the south (or as at B, fig. 47) would watch her as she traversed a higher path, as $b\ b'$, fig. 49. By timing her they could tell how long

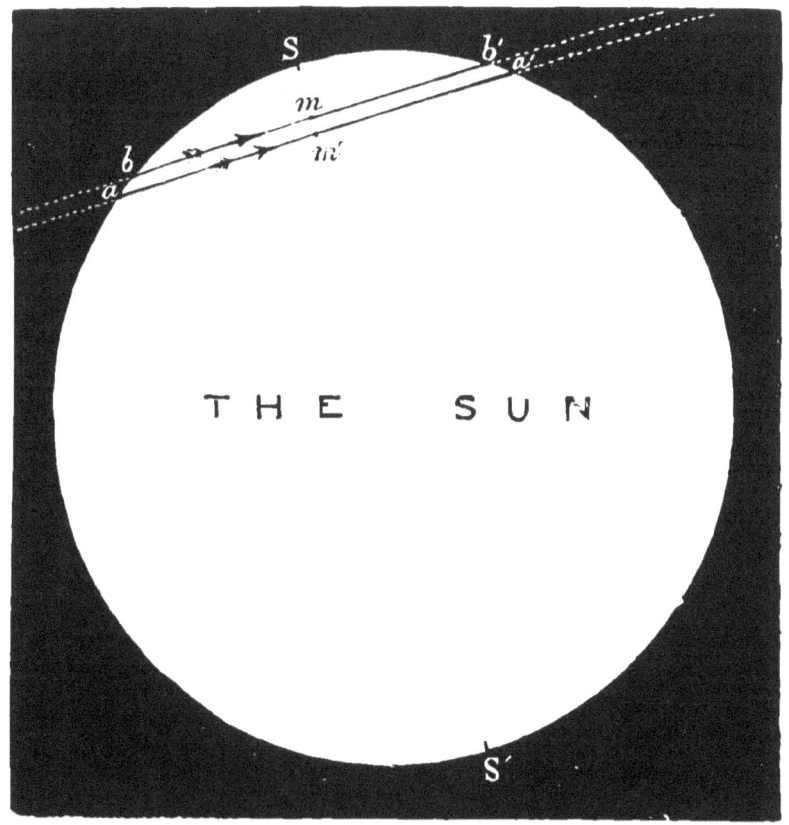

Fig. 49.

these paths were, and therefore how placed on the sun's face, as in fig. 49; that is, how far apart, which is the same thing as determining $b\ a$, fig. 48. This was Halley's plan, and as it requires that the duration of the transit should be timed, it is called the method of durations. Delisle proposed another method—viz., that one observer should time the exact moment when Venus, seen from one station, *began* to traverse the path $a\,a'$, while another should time the exact moment when she *began* to traverse the path $b\,b'$; this would show how much b is in advance of a, and thence the position of the two paths can be determined. *Or* two observers might note the *end* of the transit, thus finding how much a' is in advance of b'. This is Delisle's method, and it has this advantage over Halley's—that an observer is only required to see *either* the beginning or the end of the transit, not *both*.

I shall not here consider, except in a general way, the various astronomical conditions which affect the application of these two methods. Of course, all the time that a transit lasts, the earth is turning on her axis; and as a transit may last as long as eight hours, and generally lasts from four to six hours, it is clear that the face of the earth turned towards the sun must change considerably between the beginning and end of a transit. So that Halley's method, which requires that the whole duration of a transit should be seen, is hampered with the difficulty arising from the fact that a station exceedingly well placed for observing the beginning of the transit might be very ill placed for observing the end, and *vice versâ*.

Delisle's method is free from this objection, because an observer has only to note the beginning *or* the end, not both. But it is hampered by another. Two observers who employ Halley's method have each of them only to consider how long the passage of Venus over the sun's face *lasts;* and they are so free from all occasion to know the exact time *at* which the transit begins and ends, that theoretically each observer might use such an instrument as a stop-watch, setting it going (right or wrong as to the time it showed) when the transit began, and stopping it when the transit was over. But for Delisle's method this rough-and-ready method would not serve. The two observers have to compare the two moments at which they severally saw the transit begin,— and to do this, being many thousand miles apart, they must know the exact time. Suppose they each had a chronometer which had originally been set to Greenwich time, and which, being excellently constructed and carefully watched, might be trusted to show exact Greenwich time, even though several months had elapsed since it was set. Then all the requirements of the method would be quite as well satisfied as those of the other method would be if the stop-watches just spoken of went at a perfectly true rate during the hours that the transit lasted. But it is one thing to construct a time-measure which will not lose or gain a few seconds in a few hours, and quite another to construct one which will not lose or gain a few seconds in a journey of many thousand miles, followed perhaps by two or three months' stay at the selected station. An error of five seconds would be perfectly fatal in applying Delisle's method, and no chronometer could be trusted under the condi-

Fig. 50.

tions described to show true time within ten or twelve seconds. Hence astronomers had to provide for other methods of getting true time (say Greenwich time) than the use of chronometers; and on the accuracy of these astronomical methods of getting true time depended the successful use of Delisle's method.

Then another difficulty had to be considered, which affected both methods. It was agreed by both Halley and Delisle that the proper moment to time the beginning or end of transit was the instant when Venus was just within the sun's disc, as in fig. 50, either having just

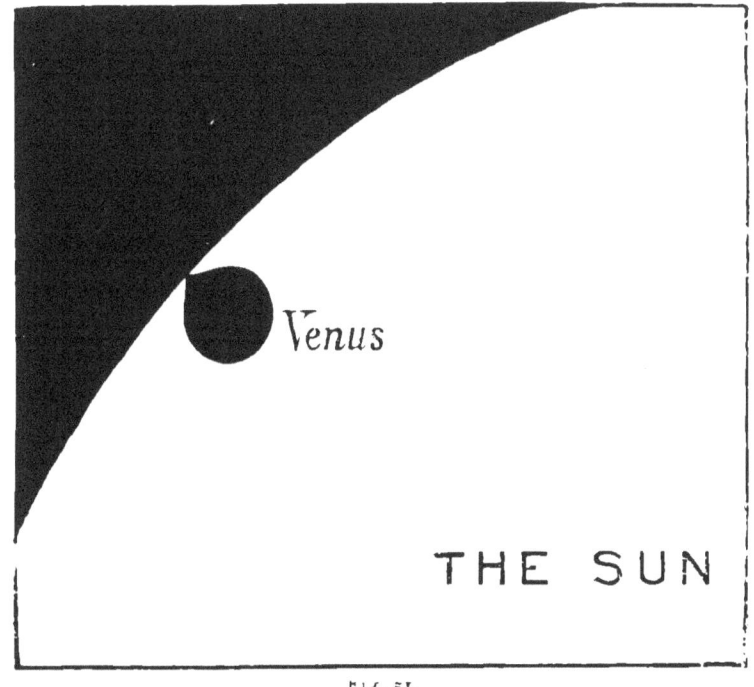

Fig. 51.

completed her entry, or being just about to begin to pass off the sun's face. If at this moment Venus presented a neatly defined round disc, exactly touching the edge of the sun, also neatly defined, this plan would be perfect. At the very instant when the contact ceased at the entry of Venus, the sun's light would break through between the edges of the two discs, and the observer would only have to note that instant; while, when Venus was leaving the sun, he would only have to notice the instant when the fine thread of light was suddenly divided by a dark point. But unfortunately Venus does not behave in

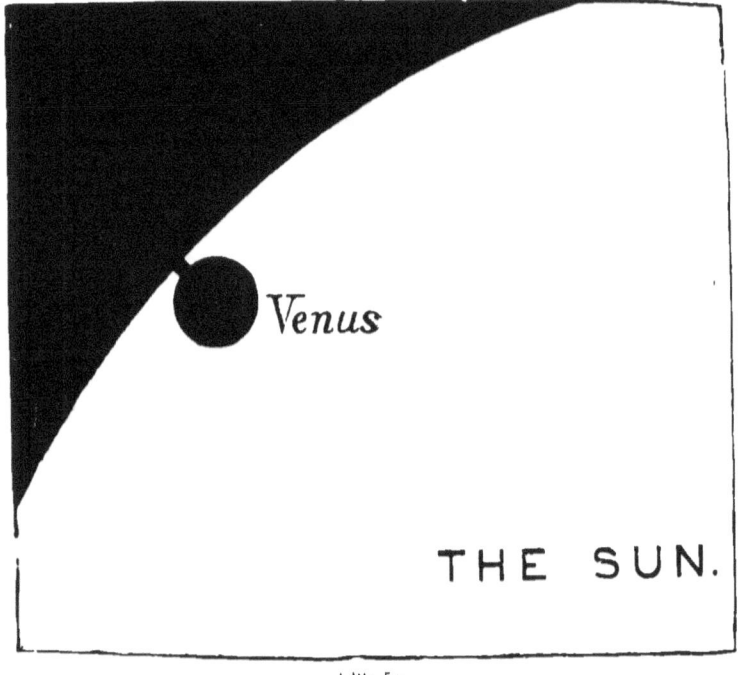

Fig. 52.

this way, at least not always. With a very powerful and very excellent telescope, in perfectly calm, clear weather, and with the sun high above the horizon, she probably behaves much as Halley and Delisle expected. But under less favourable conditions, she presents at the moment of entry or exit some such appearance as is shown in figures 51, 52, and 53, while with a very low sun she assumes all sorts of shapes, continually changing, being for one moment, perhaps, as in one or other of figs. 51, 52, and 53, and in the next distorted into some such pleasing shape as is pictured in fig. 54.

Accordingly, many astronomers are disposed to regard

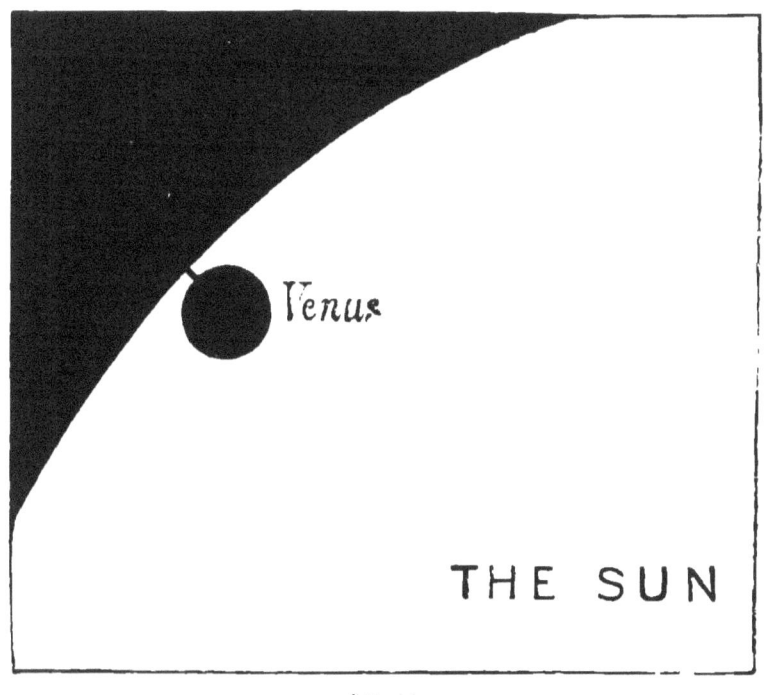

Fig. 53.

both Halley's method and Delisle's as obsolete, and to place reliance on the simple method of direct observation first described. They would, however, of course bring to their aid all the ingenious devices of modern astronomical observation in order to overcome the difficulties inherent in that method. One of the contrivances naturally suggested to meet such difficulties is to photograph the sun with Venus upon his face. The American astronomers, in particular, consider that the photographic results obtained during the transit of 1874 will outweigh those obtained by all the other methods. The German and

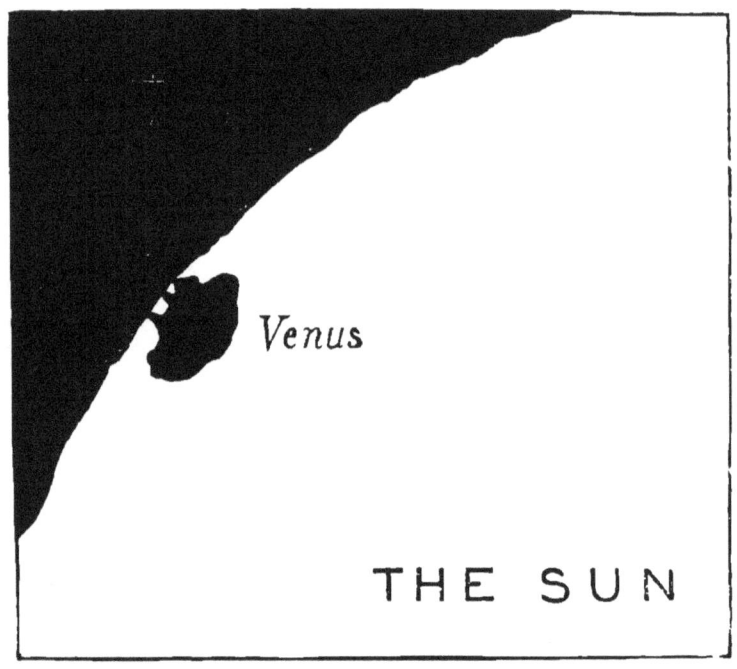

Russian astronomers, as well as those of Lord Lindsay's expedition, while placing great reliance on photography, employed also a method of measuring the position of Venus on the sun's disc, by means of a kind of telescope specially constructed for such work, the peculiarities of which need not be here considered.

The observations made in 1769 were so imperfect that astronomers deduced a distance fully 3,000,000 miles too great. Of late, other methods of observation had set them much nearer the true distance, which has been judged to lie certainly between 91,800,000 miles and 92,600,000 miles—a tolerably wide range.

But it may perhaps occur to some that the distance of the sun may be changing. The earth might be drawing steadily in towards the sun, and so all our measurements might be deceptive. Nay, the painful thought might present itself that when the observations of 1769 were made, the sun really was farther away than at present by more than 3,000,000 of miles. If this were so, the earth would, in the course of a century, have reduced her distance by fully one-thirtieth part, so that, supposing the approach to continue, she would in 3,000 years fall into the sun, while, long before that period had elapsed, the increased heat to which she would be exposed would render life impossible.

Fortunately, we know quite certainly that no such approach is taking place. It is known that the distance of the earth from the sun cannot change without a corresponding change in her period of revolution—that is, in the length of the year. The law connecting these two (indicated in the note, page 279) is such that, on the reduction of the distance by any moderate portion the period would be reduced by a portion half as great again. For instance: if the distance of the earth from the sun were reduced by a thirtieth part (or about 3,000,000 miles) the length of the year would be reduced by a thirtieth and half a thirtieth—that is, by a twentieth part, or by more than eighteen days. We know that no such change has taken place during the last century, or since the beginning of history. Nay, from the Chaldean estimate of the length of the year, which only exceeded ours by about two minutes, it is easily shown that the distance of the earth from the sun has not diminished 200 miles within the last 2,500 years. So that, assuming even that the earth is approaching the sun at this rate, or eight miles in a century, it would be 1,250,000 years before the distance would be diminished by 100,000 miles, which is the probable limit of error in the determination of the sun's distance.

If, finally, it be asked, What, after all, is the use of determining the sun's distance? the answer we shall give must depend on the answer given to the question, What, after all, is the use of knowing any facts in astronomy other than those useful in navigation, surveying, and so on? And I think that this question would introduce another and a wider one—viz., What is the use of that quality in man's nature which makes him seek after knowledge for its own sake? I certainly do not propose to consider this question, nor do I think that the reader will find any difficulty in understanding *why* I do not. But accepting the facts: (1) that we *are* so constituted as to seek after knowledge; and (2) that knowledge about the celestial orbs is interesting to us, quite apart from the use of such knowledge in navigation and surveying, it is easy to show that the determination of the sun's distance is a matter full of interest. For on our estimate of the sun's distance depend our ideas as to the scale, not only of the solar system, but of the whole of the visible universe. The size of the sun, his mass, and therefore his might, the scale of those wonderful operations which we know to be taking place upon, and within, and around the sun; all these relations, as well as our estimate of the size and mass of every planet, and therefore our estimate of the earth's relative importance in the solar system, depend absolutely and directly on the estimate we form of the sun's distance. Such being the case (this being in point of fact the cardinal problem of dimensional astronomy) it cannot but be thought that, great as were the trouble and expense of the expeditions sent out to observe the transit of 1874, they were devoted to an altogether worthy cause.

www.ingramcontent.com/pod-product-compliance
Lightning Source LLC
Chambersburg PA
CBHW021623250426
43672CB00037B/1379